Essentials of Higher Physics

Mary Webster, *B.Sc.*

Senior Lecturer in Physics, Clydebank College

Second Edition

Heinemann Educational Books
London and Edinburgh

Heinemann Educational Books
22 Bedford Square, London WC1B 3HH

LONDON   EDINBURGH   MELBOURNE   AUCKLAND
HONG KONG   SINGAPORE   KUALA LUMPUR   NEW DELHI
IBADAN   NAIROBI   JOHANNESBURG
EXETER (NH)   KINGSTON   PORT OF SPAIN

Cover photograph: Ultrafast Optical Processes. Spectral resolution of a continuum picosecond pulse. (Courtesy of Dr. P. M. Renzepis, Bell Laboratory, U.S.A.)

British Library Cataloguing in Publication Data

Webster, Mary
    Essentials of higher physics. — 2nd ed.
    1. Physics
    I. Title
    530   QC23

    ISBN 0–435–68837–5

Set in 10/11 pt Monophoto Times by
Advanced Filmsetters (Glasgow) Ltd
Printed and bound in Great Britain by
Richard Clay (The Chaucer Press) Ltd,
Bungay, Suffolk

Preface to the Second Edition

The second edition of this text contains all of the compulsory material needed for the revised SCE Higher Grade Physics syllabus. To accomplish these revisions, a certain re-ordering has been necessary together with some additions and deletions. A final Chapter has been added on Interpretation of Experimental Data.

Descriptions of experiments have been enhanced and supplemented. It is expected, however, that students will have the opportunity of performing themselves all of the experiments specified in the syllabus. As in the first edition, attention is drawn to the need for continual practice in problem solving. To this end, additional worked examples and problems have been incorporated throughout.

The revised syllabus includes five optional Special Topics, from which candidates are expected to study at least one topic. An adequate treatment of all of these topics as presently proposed and those which could occur in the future, is beyond the scope of this book. However, where appropriate the text has been extended to introduce some of these options.

Finally, I should like to thank all of those teachers who kindly have written to offer constructive comment and criticism. The second edition has been much improved by their advice.

1983   M.W.

Preface to First Edition

This book contains a succinct and cogent coverage of the material required for the Scottish Higher Physics examination.

The uses of this book are twofold: first, it can be used throughout the year as a précis of the material required for each topic. The text could be employed directly by teachers in order to alleviate the problem of providing each student with satisfactory notes. Accordingly more time could then be devoted to experiments, discussions, and problems. Secondly, it provides the basis for constructive revision prior to an examination. An adequate summary of principles, equations, and formulae is given in the framed sections. A useful reminder of important details may be obtained by reading all sentences commencing '*Note:*'.

Problems are given at the end of each lettered section in order that skill may be achieved in modes of solution and to highlight further certain aspects of the section. Detailed solutions are provided at the end of the book. Naturally these problems are in no way to be regarded as a substitute for questions from past examination papers. To accentuate the need for continual practice in problem solving an exercise section with brief answers is included.

SI units are employed throughout.

I should like to thank all those who have helped in the preparation of this book; especially certain students for their pertinent questions, my colleague Mr. George Maxwell, and the publishers for their assistance. I am particularly grateful to Mr. J. L. Patterson for so carefully reading and editing the manuscript and providing many excellent suggestions and improvements. Finally, a special note of gratitude is due to my husband, Brian Webster, for his continual encouragement.

1978   M.W.

Contents

1 Mechanics

A: Units, Dimensions, Scalars, Vectors, and Measurement

1:A.1 *SI System of Units*

When any quantity is measured its **unit** must be stated. A length of 17 has no meaning. For a unit, such as the metre, to be accepted and understood by scientists throughout the world a standard 'metre' had to be agreed upon. In Paris, in 1960, an International Committee proposed the SI system of units (Système International d'Unites), which has seven **basic units** corresponding to seven **independent** physical quantities.

Quantity	length	mass	time	electric current	thermodynamic temperature	amount of substance
Unit	metre	kilogram	second	ampere	kelvin	mole

(The seventh basic unit, the **candela**, is the unit of luminous intensity and does not require discussion in this book.)

These basic units are defined in terms of a particular property of a given substance. As examples, the definitions of the units of length, mass, and time are given below.

Length: metre
This used to be based on the length of a platinum–iridium rod kept in Paris. It is now defined in terms of the wavelength of a particular spectral line of krypton, ^{86}Kr.

$$1 \text{ metre} = 1\,650\,763.73 \text{ wavelengths of this radiation}$$

Mass: kilogram
Defined as the mass of a piece of platinum–iridium kept under standard conditions at Sèvres, near Paris, France.

Time: second
This used to be based on the mean solar day. It is now defined in terms of the period of a particular radiation emitted by caesium, ^{133}Cs.

$$1 \text{ second} = 9\,192\,631\,770 \text{ periods}$$

The **ampere** is defined in terms of the current required by two specified conductors to produce a certain force (see Section 2:B.1).

For interest it may be noted that the **kelvin** is defined in terms of the temperature at which steam, water, and ice may co-exist, this point being termed the triple point of water. Then a temperature scale based on the two fixed points for melting ice and boiling water may be established.

The **mole** is discussed later (see Section 6:B.1).*

Basic units used in this text, with their abbreviations:

metre (m)   kilogram (kg)   second (s)
ampere (A)   kelvin (K)   mole (mol)

Units for other quantities are termed **derived units**, as they may be obtained by a simple combination of the above units, *without numerical* factors.

Examples: A velocity of one metre per second is that velocity possessed by an object when its displacement increases by one metre in every second. Hence the unit for velocity is $m\ s^{-1}$.

A unit of energy would be $kg\ m^2\ s^{-2}$ but for convenience this is abbreviated to the joule, J.

The specific heat capacity of a substance has a unit of $J\ kg^{-1}\ K^{-1}$.

In terms of basic units, potential difference would have a unit of $kg\ m^2\ s^{-3}\ A^{-1}$ which is called the volt, V.

Common prefixes

$m—10^{-3}$ (milli) $\mu—10^{-6}$ (micro) $n—10^{-9}$ (nano)
$p—10^{-12}$ (pico) $k—10^3$ (kilo) $M—10^6$ (mega)

1:A.2 *Dimensions*

The dimension of a quantity is an algebraic symbol assigned to a quantity independent of its units.

Example: The distance between stars in light years; the wavelength of light in nm; the height of a man in metres; these are all length quantities with dimensions [L].

For mechanics and electricity there are four essential dimensions:

Length [L],   Mass [M],   Time [T],   Current [I].

All other mechanical and electrical quantities may be expressed in terms of these four dimensions.

However, if only one consistent set of units, the SI units, is used, it is permissible to analyse the *units* of the physical quantities under discussion. In the above example all these quantities could have been expressed in metres, i.e. all quantities of dimension [L] can have the unit metre, m.

Example: Which of the following are dimensionally equivalent: force F, kinetic energy E_k, the product of pressure and volume pV, and potential difference p.d.?

The derived unit (or units) of each quantity (or product) is expressed in terms of the basic units.

* Memorization of the details of the SI system of units is not necessary.

Quantity	Derived unit	Comment	Basic units
F	N	using $F = ma$	$\mathrm{kg\,m\,s^{-2}}$
E_k	J	$E_k = \frac{1}{2}mv^2$	$\mathrm{kg\,m^2\,s^{-2}}$
pV	$\mathrm{Pa\,m^3}$	$p = \dfrac{\text{force}}{\text{area}}$	$\mathrm{kg\,m^2\,s^{-2}}$
	(J)	$\mathrm{Pa} = \mathrm{N\,m^{-2}}; V \text{ in } \mathrm{m^3}$	
p.d.	V	$\text{p.d.} = \dfrac{\text{energy}}{\text{charge}}$ $V = \mathrm{J\,A^{-1}\,s^{-1}}$	$\mathrm{kg\,m^2\,s^{-3}\,A^{-1}}$

Thus E_k and the product pV are dimensionally equivalent and have the same unit $\mathrm{J\,(kg\,m^2\,s^{-2})}$.

Note: A ratio of two similar quantities will have *no* dimensions. For example $\sin\theta$, linear magnification, and efficiency are dimensionless.

1:A.3 *Scalars and Vectors*

A quantity is a **scalar** if it has magnitude only.

Example: The length of a piece of paper; the energy used by a light bulb.

A quantity is a **vector** if it has magnitude and direction.

Example: The force of gravity on a book lying on a table is downwards; the momentum of a moving car is forwards (unless in reverse gear!); an electric field will send electrons in a certain direction only.

Addition of scalars

Two scalar quantities may be added or subtracted arithmetically provided the units are the same.

Example: $3\,\text{cm} + 10\,\text{cm} \quad = 13\,\text{cm}$
but $2\,\text{inches} + 7\,\text{cm} = (2 \times 2.54) + 7\,\text{cm}$

Addition of vectors

With vector quantities account must be taken of their *directions*.

Example: Two forces of 4 N and 2 N are pushing an object. The angle between the two forces is 60°, see Figure 1.1(a). Determine the total force.
   The direction of the forces are shown by the arrows. The vector sum of these forces may be obtained by construction, as in Figure 1.1(b). OA is 2 units long and AB is 4 units long. Notice the position of the 60° angle.
   The resultant total force, or vector sum, is OB which may be measured. The angle AÔB will give its direction.

$$\text{OB} = 5.3\text{ units} \qquad \text{AÔB} = 41°$$

   Thus the resultant force has a magnitude of 5.3 N and acts in a direction of 41° to the 2 N force.

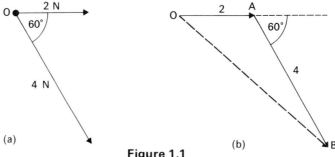

Figure 1.1

Note: The arrow on AB follows on from the arrow on OA. They must *not* be in opposing directions. (The head to tail rule.)

The same process may be applied for a number of vectors.

Example: Three forces **a**, **b**, and **c** act on a body at a point O (Figure 1.2(a)).

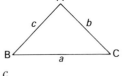

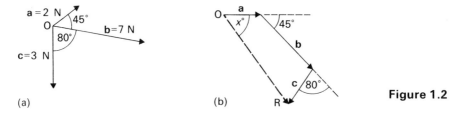

Figure 1.2

The resultant force, or vector sum, is obtained by construction (Figure 1.2(b)). The magnitude is given by OR and acts in a direction $x°$ from **a**.

Resultant force = 9 N at an angle of 55° from **a**.

In some simple cases it is easy to calculate the vector sum, particularly when right angles are involved. For the more mathematically inclined the cosine and

sine rules can be used for any triangle namely;

$$a^2 = b^2 + c^2 - 2bc \cos A \quad \text{and} \quad \frac{a}{\sin A} = \frac{b}{\sin B} = \frac{c}{\sin C}.$$

Components of a vector

Two vectors may be added together to give a single resultant vector. *Conversely,* a single vector may be split into two vectors which equal that single vector. These two vectors are usually chosen to be at right angles to each other.

The vector sum of **a** + **b** is equal to the vector **p**, see Figure 1.3(b).

$$\overrightarrow{OB} + \overrightarrow{BP} = \overrightarrow{OP} \quad \overrightarrow{BP} = \mathbf{a}$$
$$\mathbf{b} + \mathbf{a} = \mathbf{p}$$

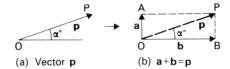

Figure 1.3 Components of a vector

a and **b** are the right angled **components** of vector **p**.
p is said to have been **resolved** into the two components **a** and **b**.

$$\text{Magnitude of } \mathbf{b} = \text{OP} \cos \alpha$$
$$\text{Magnitude of } \mathbf{a} = \text{OP} \sin \alpha \qquad A\hat{P}O = \alpha$$

Example: What are the horizontal and vertical components of a 20 N force acting at 30°
to the horizontal (Figure 1.4)?

Magnitude of the horizontal component, $\mathbf{b} = 20\cos 30° = 10\sqrt{3}\,\text{N}$
Magnitude of the vertical component, $\mathbf{a} \quad = 20\sin 30° = 10\,\text{N}$

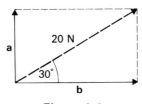

Figure 1.4

1:A.4 *Measurement*

In any experimental determination or demonstration the factors affecting the
accuracy of the measurement should be considered. These include:

(1) Limitations of the instrument in use: an ammeter of 0–10 A range with 1 A
graduations can only give readings to about the nearest 0.2 A.

(2) Choice of instrument or design of the experiment: a stop watch is a poor
device for timing a weight dropped from a table to the floor!

(3) Personal errors: starting and stopping a stop watch.

For every physical quantity encountered the following should be known:

(1) Its common unit.
(2) If it is a scalar or a vector.
(3) Its mode of measurement.

Problem
1.1 For each of the following quantities reduce their unit to the basic units of m, kg, s,
and A, and hence state their dimensions in terms of [L], [M], [T], and [I]: force,
power, frequency, half-life of a radioactive isotope, electric charge, pressure.

B: Time, Velocity, and Acceleration

1:B.1   Time

Unit: second (s), scalar.

Measured by a stop watch, electronic clock or scalar, ticker timer, multiple flash photography, or by 'stopping' a periodic motion with a stroboscope.
The advantages and disadvantages of all these methods should be considered carefully.

Stroboscopes

Mechanical: motor- or hand-driven disc with one or more equally-spaced slits.
Light: a regularly flashing light.

The rate of revolution, or flashing, is adjusted until the periodic motion viewed, for example of water waves or a rotating handle, is stationary. The number of flashes per second, or the rate of revolution R for a single slit disc, will equal the frequency of the motion viewed. For a stroboscope with N slits, the frequency is N times R, as the motion or object will be seen as each slit takes the place of the previous one.
If the rate of revolution or flashing is doubled (or tripled) then the motion or object will be seen twice (or three times). This is called double (or triple) viewing.
If the rate is halved then single viewing is still obtained but the motion or object is only viewed every other cycle. Thus the correct frequency for the motion is the *highest single-viewing frequency*.
In general if the rate of revolution is R, of an N slit strobe, and the frequency of the motion under view is f, the number of views n observed will be given by:

$$nf = RN$$

When n appears to be a fraction, this implies that the effective rate of revolution of the strobe $R \times N$ is *less* than the highest single viewing frequency, and only a fraction of the total number of cycles are being observed.

Example: A system vibrating at 20 Hz is observed through a 5 slit strobe. What is seen when the strobe rotates at (*a*) 8 Hz, (*b*) 4 Hz, (*c*) 2 Hz?
Thus: $f = 20$ Hz and $N = 5$, using $nf = RN$ for each:

(*a*) $n \times 20 = 8 \times 5$ $n = 2$ double viewing is observed.
(*b*) $n \times 20 = 4 \times 5$ $n = 1$ single viewing is observed.
(*c*) $n \times 20 = 2 \times 5$ $n = \frac{1}{2}$ single viewing is observed, but in this last case the system is observed every *other* cycle only.

1:B.2   Distance, Displacement, Speed, and Velocity

Distance

Unit: metre (m), scalar.

Distance is the total length along a specified path.

Example: A bird flies 6 km east then 3 km north. Its total flight distance is 9 km.

Displacement

Unit: metre (m), vector.

The displacement of an object is the length and direction of a line drawn from a starting point to the final position of that object.

Example: A man walks from A to B by the path shown in Figure 1.5. If the length of AB is 3.5 km then the displacement from A is 3.5 km, 78° E of N.

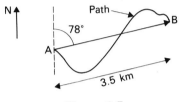

Figure 1.5

Example: A man travels 20 km north, 6 km east, and 12 km north-east. What is his distance and displacement from his starting point?

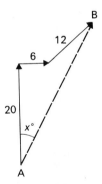

Figure 1.6

His displacement is the length AB at an angle $x°$ E of N (Figure 1.6). By construction these may be measured, giving his displacement as 32 km, 27° E of N, a three figure bearing of (027°). Remember that three figure bearings are measured clockwise from north.

His distance is $20+6+12 = 38$ km.

Speed

Unit: $m s^{-1}$, scalar.
Speed is the rate at which an object moves.

$$\text{Speed} = \frac{\text{distance covered}}{\text{time taken}}$$

$$\text{Average speed} = \frac{\text{total distance}}{\text{total time}}$$

Velocity

Unit: m s^{-1}, vector.
Velocity is the rate at which an object moves in a certain direction.

For a *constant velocity*:

$$\text{Velocity} = \frac{\text{displacement}}{\text{time taken}}$$

When the *velocity is changing*:

$$\text{Velocity} = \frac{\text{small displacement}}{\text{time taken for that small displacement}}$$

which can be written:
$$\text{velocity} = \frac{\Delta s}{\Delta t}$$

where Δ (delta) indicates a *small* change.

Ideally the time interval should approach zero.

Both speed and velocity may be measured by determining the time taken for a given distance or displacement to be covered.

Note: For an object moving in a *straight line* in one direction only, the value of the speed and velocity are *identical.*

Experiment 1.1 *To measure the speed of an object, when that speed is changing*

When the speed is changing, the time interval must be small and the average speed over this time interval measured. With a light beam/photocell and scaler arrangement the object whose speed is to be determined interrupts a narrow light beam directed onto a photocell. The electric clock or scaler connected to the photocell only records the time when the light does *not* reach the photocell. This arrangement is often called a *light gate*. Thus, if a clock records a time of interruption of 0.36 s, when an object of diameter (or length) 2.7 cm passes through the light beam, the speed of the object is $\frac{2.7}{0.36} = 0.75\,\text{cm s}^{-1}$.

With a ticker timer operating at 50 Hz, the time interval is 0.02 s between dots, and the distance can be measured directly off the tape, but the drag or friction on the tape of a ticker timer limits the accuracy of its use for determining speed or velocity. Multiple flash or strobe photography avoids this error. Again the time interval is the reciprocal of the flash rate but a metre rule or scale must be included in the photograph in order to measure distances from the photograph, which is not full size.

For a *uniformly changing* velocity and *linear* motion.

$$\text{average velocity} = \tfrac{1}{2} \text{ (initial velocity + final velocity)}$$

Note: The difference between speed and velocity: speed is a scalar quantity with magnitude only; a negative velocity implies a reverse direction.

Relative velocity

To determine the relative velocity of a moving object A with respect to another moving object B, the velocity of B must be subtracted from that of A. (This is equivalent to bringing B to rest by subtracting the velocity of B from both A and B.)

Example: An object A travels due north at $10\,\text{m s}^{-1}$ and passes another object B also moving north at $6\,\text{m s}^{-1}$. The velocity of A relative to B is $(10-6) = 4\,\text{m s}^{-1}$ due north. Notice that the velocity of B relative to A is $4\,\text{m s}^{-1}$ due south, i.e. $-4\,\text{m s}^{-1}$ north, see Figure 1.7(b).

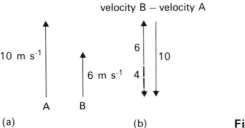

(a)   (b)   **Figure 1.7**

Sometimes the objects are travelling in different directions.

Example: What is the relative velocity of a wind blowing due west at $30\,\text{m s}^{-1}$ to a person standing on a ship moving north at $40\,\text{m s}^{-1}$?
The velocity of the ship must be subtracted (vectorially) from the velocity of the wind.
Subtraction of a vector $40\,\text{m s}^{-1}$ north is equivalent to adding a vector of $40\,\text{m s}^{-1}$ south, see Figure 1.8(b).

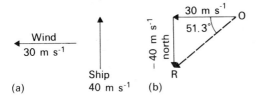

(a)   40 m s⁻¹ (b)   **Figure 1.8**

To a person on the ship the wind appears to have a velocity of magnitude OR = $50\,\text{m s}^{-1}$ in a direction 53.1° S of W.

A common problem in physics examinations concerns two objects moving relative to a third.

Example: A man walks across the deck of a ship at $5\,\text{m}\,\text{s}^{-1}$ due west with respect to the ship. The ship is moving at $12\,\text{m}\,\text{s}^{-1}$ north with respect to the sea. Determine the man's velocity relative to the *sea*.

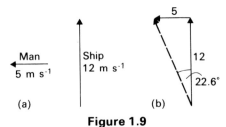

Man
5 m s⁻¹
(a)

Ship
12 m s⁻¹
(b)

5

12

22.6°

Figure 1.9

Here the *sum* of these two velocities is required, namely $13\,\text{m}\,\text{s}^{-1}$ in a direction 22.6° W of N, see Figure 1.9(b).

1:B.3 *Acceleration*

Unit: $\text{m}\,\text{s}^{-2}$, vector.
Acceleration is the rate at which the velocity is changing.

For a *uniform acceleration*:

$$\text{Acceleration} = \frac{\text{change in velocity}}{\text{time taken for that change}}$$

For a *non-uniform acceleration* the time interval should be as small as possible, tending to zero.

Note: The acceleration should be stressed as being the change in velocity with *time* and not with *distance*. It is the change in velocity which takes place in one second.

Acceleration is a vector quantity, and the *sign* of the acceleration is determined by the *change* in velocity.

Equations of motion

For a *uniform acceleration* and *linear motion:*

u—initial velocity $\qquad v$—final velocity
a—acceleration $\qquad t$—time taken
s—displacement

$$a = \frac{v - u}{t}$$

which is $\qquad\qquad v = u + at \qquad\qquad\qquad (1)$

For a uniform acceleration:

$$\text{Average velocity} = \tfrac{1}{2}(u + v) = \frac{s}{t}$$

which together give

$$s = \tfrac{1}{2}t(u+v)$$

Using $v = u + at$ and substituting for v,

$$s = \tfrac{1}{2}t(u+u+at) \qquad\qquad s = ut + \tfrac{1}{2}at^2 \tag{2}$$

Eliminating t from (1) and (2) gives

$$v^2 = u^2 + 2as \tag{3}$$

These three equations are called the **equations of motion**.

$$v = u + at \qquad s = ut + \tfrac{1}{2}at^2 \qquad v^2 = u^2 + 2as$$

and providing

(1) the acceleration is uniform and the motion is linear,
(2) the units are correct, and
(3) the acceleration is given a negative sign for a deceleration,

any unknown quantity may be calculated from these three equations.

Note: The equations apply to the *vector* quantities s, u, v. These are given negative signs for movement in the reverse direction. In projectile problems it is advisable to state clearly whether the positive direction is upwards or downwards.

Example: List the quantities given in the question below with their units and add to the list the quantity to be calculated. Then choose that equation containing those four quantities.
  A ball accelerates at $5\,\mathrm{m\,s^{-2}}$ from rest for $4\,\mathrm{s}$. What distance has it travelled?

$a = +5\,\mathrm{m\,s^{-2}}$ (There is no information about the final velocity nor is its
$u = 0$ calculation required so it is not included in the list.)
$t = 4\,\mathrm{s}$
$s = ?$

Check that the units are consistent and a has the correct sign.

The equation required is $s = ut + \tfrac{1}{2}at^2$.
$$s = 0 + \tfrac{1}{2} \times 5 \times 16$$
Distance, $s = 40\,\mathrm{m}$

1:B.4 *Graphs of Acceleration a, Velocity v, Displacement s, with Time t*

Interpretation of graphs

When two quantities are plotted against each other the **gradient** (or slope) of the graph obtained is the rate at which one quantity (usually the ordinate y) varies with respect to the other (usually the abscissa x).

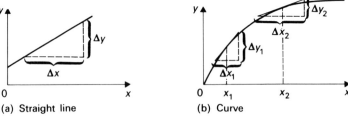

(a) Straight line   (b) Curve

Figure 1.10   Gradients

The gradient is the ratio $\dfrac{\Delta y}{\Delta x}$.

Note: If the graph is a straight line the gradient $\dfrac{\Delta y}{\Delta x}$ is constant. If the graph is a curve the tangent to the curve at the point of interest is drawn and the gradient $\dfrac{\Delta y}{\Delta x}$ determined for that point. In Figure 1.10(b), the gradient at x_2 is *less* than that at x_1, showing that the gradient is decreasing.

Graphs for linear motion

Displacement, velocity, and acceleration are *vector* quantities with magnitude and *direction*. The graphs below illustrate movement in the forward, *positive*, direction only.

(1) **No acceleration:** *constant velocity.*

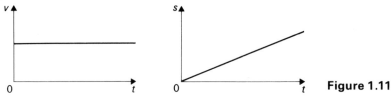

Figure 1.11

(2) **Constant acceleration:** *motion in the forward, positive, direction.*

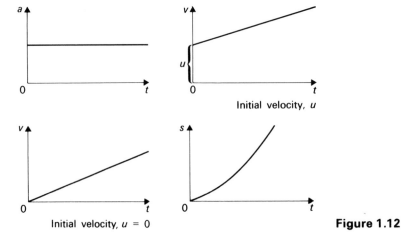

Initial velocity, u

Initial velocity, $u = 0$

Figure 1.12

(3) **Constant deceleration:** *motion in the forward, positive, direction.*

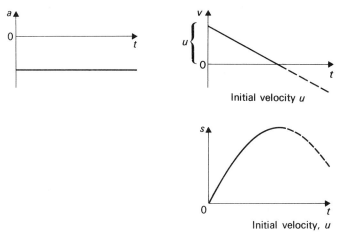

Figure 1.13

Observe that the v/t graph will cross the x axis. The negative velocity implies motion in the *reverse* direction. This is further illustrated by the s/t graph. The displacement reaches a maximum, then it will decrease as the object returns to its starting point. Also, observe that a 'speeding-up' in the reverse direction would still give a negative value for a as shown in the a/t graph. The gradient of the v/t graph *remains* negative when the graph crosses the axis.

(4) **Changing acceleration and deceleration**

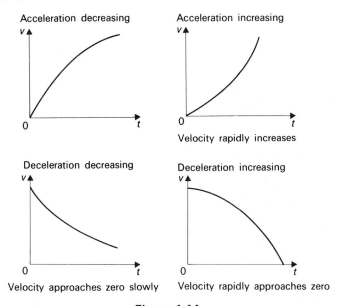

Figure 1.14

If a steel ball is dropped into a viscous liquid the acceleration will decrease to zero, and the ball will fall with a **terminal** velocity. This is an example of the first graph, acceleration decreasing.

An experiment illustrating changing acceleration is given on page 33.

Interpretation of the graphs in Figures 1.11–1.14

(1) The gradient of a displacement/time graph gives the velocity at that time. Thus a straight-line displacement/time graph implies a constant velocity, but a *curve* implies a *changing* velocity because the gradient is changing.

(2) The gradient of a velocity/time graph gives the acceleration at that time. Again observe the straight-line velocity/time graphs where the acceleration is constant.

(3) The area under a velocity/time graph gives the displacement covered in that time interval. For example, an object increases its velocity from rest to v_1 in time t_1 with uniform acceleration.

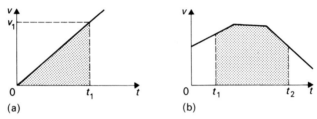

Figure 1.15 Area under a graph

The velocity/time graph will be a straight line (Figure 1.15(a)).

$$\text{Displacement} = \text{average velocity} \times \text{time} = \tfrac{1}{2}v_1 t_1$$

$$= \text{area under the graph.}$$

This can be extended to any velocity/time graph irrespective of its shape. In Figure 1.15(b), the displacement between t_1 and t_2 is equal to the shaded area.

The important points are summarized below:

> The gradient of a displacement/time graph gives the velocity.
> The gradient of a velocity/time graph gives the acceleration.
> The area under a velocity/time graph gives the displacement.

Note: The area under an acceleration/time graph gives the *change in velocity*. For a constant acceleration, the change in velocity $(v-u)$ is equal to $a \times t$. This can be extended to any a/t graph.

Example: Show, on a velocity/time graph, the areas represented by the terms ut and $\tfrac{1}{2}at^2$ in the equation $s = ut + \tfrac{1}{2}at^2$, and interpret the meanings of these terms.

In Figure 1.16 the darker shaded area ABEF $= ut$. This is the displacement which would occur if the object *only* had a steady velocity u.

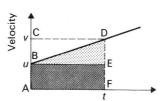

Figure 1.16

The lighter shaded area $\text{BED} = \frac{1}{2}\text{BE} \times \text{ED}$ $\qquad$ $\text{ED} = $ change in velocity
$$= \frac{1}{2}t \times at \qquad\qquad = a \times t$$
$$= \frac{1}{2}at^2$$

This area represents the displacement due to the *change* in velocity.

Example: A ball is thrown vertically upwards with a velocity of $30\,\text{m s}^{-1}$. Draw velocity/time and speed/time graphs to illustrate the motion of the ball in the first six seconds.

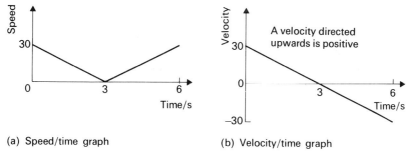

(a) Speed/time graph $\qquad\qquad\qquad$ (b) Velocity/time graph

Figure 1.17

The ball will decelerate due to gravity, (see next section). It will come to rest after 3 s at the top of its flight, then fall, accelerating under the influence of gravity, until it reaches the ground after 6 s. Notice that the velocity is negative between 3 s and 6 s since the ball is falling downwards. Upwards was chosen as the positive direction.

Note:

(1) Velocity has magnitude and direction but speed has magnitude only. However, a single velocity/time graph can only show direction for linear motion, in this case up and down vertically. Similarly a displacement/time graph may only be used for linear motion, where a negative displacement implies a reverse direction. Unless stated otherwise it is usually assumed that a motion is rectilinear.

(2) For any particular problem, one direction must be chosen as the forward, positive, direction giving the reverse direction as negative. Since velocity is *vector*, velocities in the *reverse* direction are *negative*. Thus an acceleration, a 'speeding up', in the reverse direction will be negative. But, a deceleration, a 'slowing down', in the reverse direction will be positive.

Example: State the acceleration of an object, travelling in the reverse direction, whose speed changes from $10\,\text{m s}^{-1}$ to $30\,\text{m s}^{-1}$ in 4 s.

The reverse direction implies negative values for the velocities.

$$\text{Change in velocity} = [-30-(-10)] = -20\,\text{m s}^{-1}$$
$$a = -5\,\text{m s}^{-2}; \text{ a negative value of } a.$$

Example: Determine the acceleration when the velocity of an object changes from $-11\,\text{m s}^{-1}$ to $-3\,\text{m s}^{-1}$ in 2 s.

The negative velocities indicate motion in the reverse direction.

$$a = \tfrac{1}{2}[-3-(-11)] = +4\,\text{m s}^{-2} \qquad\qquad \text{a positive value!}$$

(See Problem 1.7 page 20.)

1:B.5   *Gravity and Projectiles*

Acceleration due to gravity

The acceleration due to gravity, g, is the acceleration that any *freely falling* body will have, regardless of its mass, due to the attraction of the Earth. The value of g *varies* over the surface of the Earth. In Britain, g is usually taken as $9.8\,\text{m s}^{-2}$ or $9.8\,\text{N kg}^{-1}$ (see Section 1:C.1) but as the value given in most examination papers is $10\,\text{m s}^{-2}$, this value will be used in all examples.

Experiment 1.2   *To measure the acceleration due to gravity* g

An object which falls *freely* near the earth's surface will accelerate with a value of g.

(a) A ticker tape is attached to a steel ball and the ball is released from the edge of the bench. The acceleration of the ball is determined from the tape. The measured value of g will tend to be too low because of the drag of the ticker timer on the tape. This error will be considerably more if a lighter ball is used. Alternatively strobe photography can be used to give a more accurate result.

    The ability to calculate accelerations from ticker tapes or multiple-flash photographs should be understood. See Problem 1.11 on page 21.

(b) An electronic method can be used to obtain a measurement of the time taken t for a steel ball to fall from rest through a measured distance s.

    With the d.c. supply on, the soft iron becomes a temporary magnet holding the steel

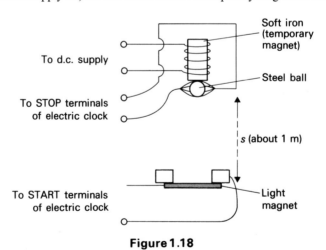

Figure 1.18

ball in position. The STOP circuit of the electronic clock is completed by wires touching either side of the ball. The lower, light, magnet completes the START circuit. When the d.c. supply is switched off, the soft iron ceases to be a magnet and the ball falls, breaking the STOP circuit. The clock commences to time. When the ball knocks away the lower magnet, the START circuit is broken and the clock ceases to time. The time of falling a distance s is displayed on the clock, and g calculated using $s = \frac{1}{2}gt^2$ since the initial velocity $u = 0$.

Alternatively the connections to the STOP terminals are incorporated in the electromagnet circuit. When the *supply* to the electromagnet is switched off, the clock commences to time. This method avoids the use of wires touching the ball, but it has the disadvantage of producing a systematic error in the time measurement due to magnetic hysteresis, that is, the soft iron does not lose its magnetism *immediately* when the supply is switched off. Hence, there is a small time delay before the ball drops down, (see Problem 8.11 page 246).

(c) A simple pendulum has a period $T = 2\pi\sqrt{\dfrac{l}{g}}$, where l is the length of the pendulum. A series of readings of T and l are taken. A graph of T^2 against l will be a straight line with a gradient of $4\pi^2/g$.

The ability to understand the methods of calculating accelerations from experimental data should be understood. Once these have been mastered, it can be interesting to link experiments to a microcomputer (or microprocessor system) in order to obtain more precise results. (See Bibliography.)

A bouncing ball

Figure 1.19 shows downward velocity as negative. The ball, released from rest at time 0, fell downwards due to gravity and its velocity increased between time 0

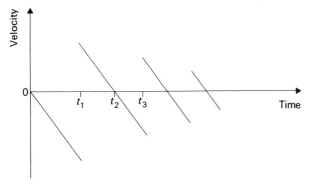

Figure 1.19

and time t_1. At t_1 the ball struck the ground and the velocity became directed upwards. The ball's upward velocity slowed between t_1 and t_2 and then increased in a downward direction between t_2 to t_3.

Note: The ball is stationary, with its velocity zero at time t_2, at the top of the bounce. Also observe the loss of maximum velocity at each bounce due to a loss of energy at impact (see Section 1:C.4).

Care must be taken when using the equations of motion to state the positive direction, and to use a, the acceleration, as $-10\,\mathrm{m\,s^{-2}}$ when an object is thrown upwards.

Horizontal and vertical velocities

Sometimes an object already has a steady horizontal velocity and falls under gravity where it acquires a changing vertical velocity. For example, an object released from an aircraft falls to the ground. The horizontal velocity will continue almost unchanged, but the vertical velocity will increase by $10\,\mathrm{m\,s^{-1}}$ in every second. The magnitude of the vertical velocity must be calculated separately. The resultant velocity at any time may be obtained by the vector addition of the two velocities.

Note: The independence of horizontal and vertical velocities.

Example: An object is projected horizontally with a speed of $4\,\mathrm{m\,s^{-1}}$ off the end of a bench 1.25 m above the ground. How far away from the base of the bench does it land?
    The horizontal component of the velocity remains at $4\,\mathrm{m\,s^{-1}}$.
To find the horizontal distance the time must be determined.
Vertically there is acceleration under gravity. Downwards is the positive direction.

$u = 0$, $s = 1.25\,\mathrm{m}$, $a = 10\,\mathrm{m\,s^{-2}}$, $v = ?$
using $s = ut + \frac{1}{2}at^2$ $1.25 = \frac{1}{2} \times 10 \times t^2$
                    $t = 0.5\,\mathrm{s}$

Using this time the horizontal distance can be calculated.

$$\text{Horizontal distance} = 4 \times 0.5 = 2\,\mathrm{m}$$

It is important to remember that the *time* taken 'vertically' is the same as the *time* taken 'horizontally', since it is only the components of the velocity in the two directions which are under consideration.
    A similar approach is used when an object is projected upwards at an angle.

Example: A ball is thrown at 30° to the horizontal with a velocity of $80\,\mathrm{m\,s^{-1}}$.
(*a*) What are the initial horizontal and vertical components of the velocity of the ball?
(*b*) What is the resultant velocity of the ball after 3 s?

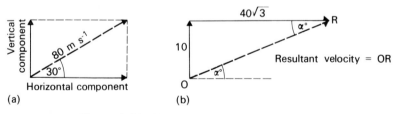

Figure 1.20 Horizontal and vertical velocities

From Figure 1.20,
(*a*) the horizontal component of the velocity $= 80\cos 30° = 40\sqrt{3}\,\mathrm{m\,s^{-1}} = 69.3\,\mathrm{m\,s^{-1}}$
    the vertical component of the velocity $= 80\sin 30° = 40\,\mathrm{m\,s^{-1}}$
(*b*) The horizontal and vertical velocities may be treated independently.
    The horizontal component of the velocity remains *unchanged* $= 69.3\,\mathrm{m\,s^{-1}}$. In the vertical direction there is a retardation of $10\,\mathrm{m\,s^{-2}}$ due to gravity.
    Let upwards be positive.
    Using $v = u + at$ where $v = ?$, $u = 40\,\mathrm{m\,s^{-1}}$, $a = -10\,\mathrm{m\,s^{-2}}$, $t = 3\,\mathrm{s}$,
    This gives $v = 10\,\mathrm{m\,s^{-1}}$.

Hence the velocity after 3 s is the vector addition of these two velocities. This may be determined by construction or calculation, as in Figure 1.20(b).

Resultant velocity $= 70 \, \text{m s}^{-1}$, at an angle α of 8.2° to the horizontal.

Uniform motion in a circle

Consider an object with uniform speed travelling in a circle. For example: a conker whirled in a horizontal circle on the end of a piece of string; a satellite moving in a circular orbit around the Earth.

Because the *direction* of motion changes with time the velocity is *not* constant and the object has an acceleration. In both of the above examples there is a force towards the centre of the circle; namely the tension in the string or the gravitational attraction of the Earth.

The direction of the velocity at any instant is tangential to the circle. The magnitude of the velocity at P equals the magnitude of the velocity at Q but since velocity is a vector quantity the *change in velocity*, Δv, must be determined by vector subtraction, as in Figure 1.21(b). This change in velocity is finite,

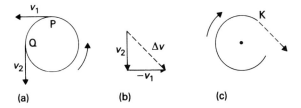

Figure 1.21   Circular motion

therefore the object must have an acceleration. Remember only the change in *speed* is zero!

If in the above example the string holding the conker is cut at K, the force on the conker becomes zero and the velocity will no longer change (see Section 1:C.2). The conker will fly off at a *tangent* to the circle at the point K, in Figure 1.21(c).

(Apart from this example, which emphasizes the vector nature of velocity, motion in a curved path will not be considered further.)

Problems
1.2   (a) The knob of a ticker timer, run off the 50 Hz mains, is illuminated by a 5-slit strobe. How many revolutions per second are required for single viewing?
(b) A water wave is viewed through a 6-slit strobe and is 'stopped' when the frequency of the strobe is 20, 30, and 60 revolutions per second, but double viewing is obtained at 120 revolutions per second. What is the frequency of the waves?

1.3   A disc with a white radial mark rotates clockwise at 60 Hz. It is observed through a four-slit motor strobe driven by a signal generator. State:
(a) the frequency of the generator needed to produce triple viewing.
(b) two possible frequencies which would produce single viewing.
(c) what would be observed if the frequency of the strobe was about 16 Hz?

1.4   A car increases its velocity from rest to $30 \, \text{m s}^{-1}$ in 2 seconds. What is its acceleration and how far has it travelled?

1.5 Describe the motion of the object whose velocity v is shown by the graph in Figure 1.22.

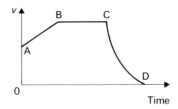

Figure 1.22

1.6 Draw displacement/time, velocity/time, and acceleration/time graphs for the following motions:
(a) an object falling from rest under the influence of gravity.
(b) a parachutist.
(c) a box dropped from a balloon travelling vertically upwards at $30\,\mathrm{m\,s^{-1}}$.
(d) a motor car starting from rest, accelerating at $12\,\mathrm{m\,s^{-2}}$ for $5\,\mathrm{s}$, travelling at constant velocity for $8\,\mathrm{s}$, then skidding to a halt in $4\,\mathrm{s}$.
Numerical values are required on the axes for (c) and (d).

1.7 The graph in Figure 1.23 shows the velocity of an object for 15 seconds.

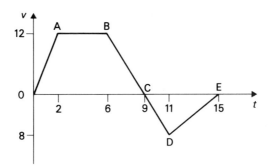

Figure 1.23

(a) Describe the **speed** over each section: OA, AB, BC, CD, DE.
(b) State the **direction** of the motion in each section.
(c) Sketch the corresponding a/t graph for the entire 15 seconds.

1.8 A train 50 m long standing at rest in a station is given the all clear by a signal 150 m in front of the engine. The train accelerates uniformly at $1\,\mathrm{m\,s^{-2}}$ out of the station. Calculate: (a) the velocity with which the front and back of the train each pass the signal. (b) The time taken for the whole train to pass the signal.

1.9 A ball is projected at an angle of 30° to the horizontal with a velocity of $40\,\mathrm{m\,s^{-1}}$ near to the edge of a cliff. The ball strikes a rock at the foot of the cliff 10 seconds later.
(a) Draw graphs of the horizontal and vertical components of the velocity for the total flight of the ball. Numerical values are required on the axes.
(b) How far away from the foot of the cliff is the rock on which the ball landed?
(c) Determine the velocity of the ball just before it landed on the rock (magnitude and direction).

1.10 An object is released from an aircraft travelling horizontally at $1000 \, \mathrm{m \, s^{-1}}$ and takes 40 s to reach the ground.

(*a*) What is the horizontal distance travelled?

(*b*) What is the height of the aircraft when the object was released?

(*c*) What is the vertical velocity before impact?

(*d*) Show by graphical means or otherwise, the magnitude of the resultant velocity just before hitting the ground.

1.11 Figure 1.24 shows the distance between successive dots on a piece of ticker tape whose dots are made at the rate of 50 per second.

(*a*) What is the velocity at the point X?

(*b*) What can be stated concerning the acceleration?

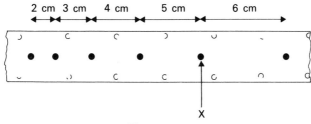

Figure 1.24

1.12

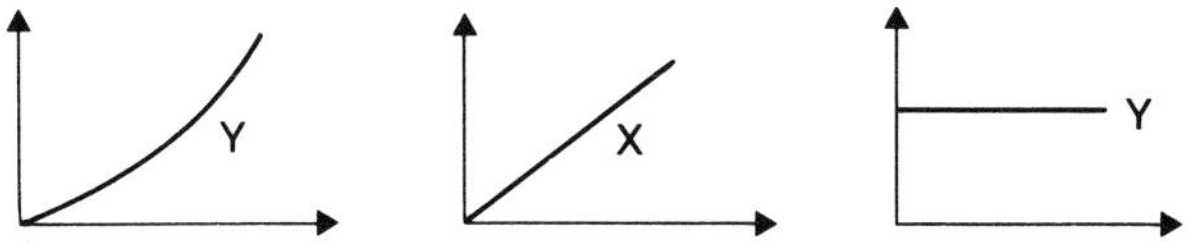

Figure 1.25

The graphs of Figure 1.25 show two types of motion, constant velocity X, and constant acceleration Y. Copy the graphs and label the axes s/t, v/t, or a/t as appropriate.

C: Mass, Force, Work, Energy, Power, and Momentum

1:C.1 *Mass, Weight, and Gravitational Field Strength*

Mass

Unit: kilogram (kg), scalar.

The mass of an object will remain unchanged unless it undergoes some chemical reaction or radioactive decay.

Mass has **inertia**, that is a resistance to motion. A mass also has *gravitational attraction* to other masses.

Experiment 1.3 *To show inertia*

The rate of vibration of a wig-wag machine and a hacksaw blade are observed to decrease when masses are added. This shows that mass has inertia, that is a desire to remain at rest.

In principle such apparatus could be calibrated with standard masses and a measurement of frequency used to obtain values of unknown masses. In practice many balances for determining mass involve a lever, often with movable standard masses to restore equilibrium.

Weight

Unit: newton (N), vector.
The weight of an object is the **force** of gravitational attraction exerted by a large body, such as the Earth, on that object. The direction of the force will be towards the centre of gravity of the large body. Thus the weight of an object is a variable quantity depending on the position of the object. The force of attraction will be greater for an object on the Earth than on the Moon.

If an object is held above the Earth's surface and *released*, it will accelerate downwards with an acceleration g, where g is the acceleration due to gravity at that place. Since the force F producing an acceleration a is given by $F = ma$ (see page 24), then the weight W of an object of mass m is thus:

$$W = mg$$

where m is the inertial mass.
Notice that as g varies over the Earth's surface, and is considerably less on the Moon, W also *varies*.

As $m = \dfrac{W}{g}$, if g is known at a given position the mass of an object may be obtained indirectly from a measurement of the weight.

Note: The difference between mass and weight: mass is scalar and remains constant with position but weight is vector and varies depending on the distribution of other masses in that region.

The gravitational field

A **field** is a concept used to describe some interaction between two objects or particles. One particle may produce a 'field' around it and the 'field' causes the second particle to experience some interaction, a force for example. A mass such as the Earth produces a **gravitational field** and another mass, the Moon, will experience a *force* of attraction due to this field. (See also electric fields, Section 2:A.2.)

Any mass will produce a gravitational field around it, but if the mass is small the field will be weak. Only large masses, such as the Sun, Earth, or Moon, will produce a sizeable gravitational field with easily *observable* effects on other masses.

Gravitational field strength

Unit: $N\,kg^{-1}$, vector.
The gravitational field strength at a point may be defined as the force on a unit mass placed at that point.

$$\text{Gravitational field strength} = \frac{F}{m} \begin{matrix} \diagup \text{N} \\ \diagdown \text{kg} \end{matrix}$$

where m is the gravitational mass.

(In the equation $F = ma$, m is the inertial mass. A force F is needed to produce an increase in speed, because of the inertia of the mass. The gravitational mass of an object will experience a force in a gravitational field. The equivalence of inertial mass and gravitational mass has been demonstrated experimentally to 1 part in 10^{11} so the symbol m will be used for *both* masses.)

An object *stationary* on the Earth's surface will experience a pull, a force, towards the Earth; this is the weight W. The relationship between the force of gravity, the weight, and the gravitational field strength on a planet is thus:

Weight = gravitational field strength × mass

But using $F = ma$ it has been shown that $W = mg$. Since inertial and gravitational masses are equal then:

gravitational field strength = acceleration due to gravity

This gives *two* units for g:

$$g = 10 \, \text{m} \, \text{s}^{-2} \quad \text{or} \quad g = 10 \, \text{N} \, \text{kg}^{-1}$$

Note: For any field the *strength* of the field will be given by the *force* on an appropriate test body. The dominant physical property of the test body will be the same as the physical property producing the field. For example, a gravitational field is produced by a large *mass*, and the gravitational field strength is the force on a unit test *mass*. An electrostatic field is produced by a *charged* object, and the electric field strength is determined by the force on a unit positive charge, (see page 45).

1:C.2 *Force, Newton's Laws of Motion, and Impulse*

Force

Unit: newton (N), vector.

If a force acts on a body it will change or tend to change the state of rest or uniform velocity of the body.

Examples: When a ball is hit with a bat, the ball acquires a velocity. A book on a table is being pulled towards the centre of the Earth but the table prevents it attaining any velocity by exerting an equal force upwards.

An object will only move when it experiences an *unbalanced* or *resultant force*. If an unbalanced force acts on a body it will either be accelerated or decelerated as its velocity changes.

Newton's First Law of Motion

A body will continue in its state of rest or uniform velocity unless acted upon by an unbalanced force.

Momentum

Unit: kg m s^{-1}, vector.
The momentum of a body at any instant is the product of the mass and the velocity at that instant. It acts in the same direction as the velocity.

$$\underset{\text{kg m s}^{-1}}{\text{Momentum}} = \underset{\text{kg}}{\text{mass}} \times \underset{\text{m s}^{-1}}{\text{velocity}}$$

Newton's Second Law of Motion

The rate of change of momentum of a body varies directly as the force causing the change and takes place in the same direction as the force.

$F \propto$ rate of change of momentum

$$\propto \frac{\text{change of momentum}}{\text{time taken for this change}}$$

$$\propto \frac{mv - mu}{t}$$

$$\propto m\left(\frac{v - u}{t}\right)$$

$F \propto ma$

u—initial velocity
v—final velocity
m—mass
a—acceleration

If the unit of force is chosen as the newton, and is *defined* as that force which gives a mass of 1 kg an acceleration of 1 m s^{-2}, then the constant of variation is unity.

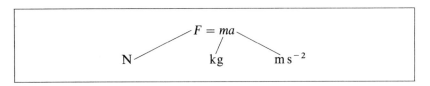

$$\underset{\text{N}}{\qquad} \overset{F = ma}{\underset{\text{kg} \qquad \text{m s}^{-2}}{\qquad}}$$

Note: An *unbalanced*, or *resultant*, force F produces an acceleration a.

Example: What force will give a mass of 4 kg an acceleration of 3 m s^{-2} if there is a constant frictional resistance of 2 N.

Force required to produce the acceleration $= 12 \text{ N}$
Total force required $= 14 \text{ N}$

The equation $F = ma$ is often used with the equations of motion to calculate unknown quantities.

Example: A 0.5 kg ball travelling at $40 \, \mathrm{m\,s^{-1}}$ hits a sandbag and comes to rest in a distance of 20 cm. What is the average resistive force acting on the ball?

$$a = ? \quad u = 40 \, \mathrm{m\,s^{-1}} \quad v = 0 \quad s = 20 \, \mathrm{cm} = 0.2 \, \mathrm{m}$$

$$\text{using } v^2 = u^2 + 2as, \qquad a = -4000 \, \mathrm{m\,s^{-2}}$$

$$\text{Resistive force} = 0.5 \times 4000 = 2000 \, \mathrm{N}$$

Experiment 1.4 *Verification of* F = ma

(1) By using a friction compensated plane, the equation $F = ma$ can be verified by applying a constant force F to a trolley of mass m, and measuring the acceleration from an attached ticker tape. The force can be provided by an elastic cord kept stretched to a certain distance. This is repeated using different numbers of stretched elastic cords, i.e. different forces, and measuring the corresponding accelerations, to show $F \propto a$. The experiment is then repeated using only one elastic cord, one force, applied to different masses to show $a \propto \dfrac{1}{m}$.

Instead of using elastic cords, the force can be provided by attaching a cord to the trolley, passing the cord over a smooth pulley at the end of a bench, and attaching a mass M, (see experimental arrangement in Figure 1.29). The force F on the trolley is then Mg.

(2) Alternatively, since the newton is defined as that force which gives a mass of 1 kg an acceleration of $1 \, \mathrm{m\,s^{-2}}$, the calibration of a newton balance can be checked. A trolley is pulled down a friction compensated slope by a newton balance held steadily extended to its 1 N mark. The acceleration can be determined by, either using a stop watch to time the trolley from rest over a distance of one metre ($s = \frac{1}{2}at^2$), or 'directly' using multiple-flash photography, or a ticker timer and tape. The product of mass and acceleration can be compared with the reading on the balance. Any difference between these two is largely accounted by the experimental error incurred in using the balance in a horizontal position, and not in the vertical position in which it was calibrated. The use of a stop watch could also introduce error.

Impulse

Unit: N s, vector.

The impulse of a force is equal to the force multiplied by the time interval over which the force acts.

From Newton's Second Law,

$$F = m \frac{\Delta v}{\Delta t} \qquad \begin{array}{l} \Delta v - \text{change in velocity} \\ \Delta t - \text{time taken} \end{array}$$

then

$$\text{Impulse} = F \, \Delta t = m \, \Delta v$$

Note: The impulse applied to an object is equivalent to the change of momentum experienced by that object. The unit N s is equivalent to $\mathrm{kg\,m\,s^{-1}}$.

Example: A stationary ball of mass 0.1 kg is hit with a hockey stick. If the time of contact is 4 ms and the average force over this time is 750 N state the velocity with which the ball leaves the stick.

Impulse $F \Delta t$ = change of momentum $m \Delta v$

$750 \times 4 \times 10^{-3} = 0.1 \times v$ initial velocity = 0

Velocity on leaving the stick $v = 30 \, \text{ms}^{-1}$

Example: An object of mass 50 g travelling at $20 \, \text{m s}^{-1}$ hits a wall and rebounds back at $10 \, \text{m s}^{-1}$. Calculate the impulse and average force on the ball if the impact lasts 0.02 s.

Since the question gives information concerning the velocities, the impulse will be calculated from the change in momentum.

The change in momentum of the ball is given by

(momentum of ball after impact) − (momentum of ball before impact)

Hence, let the positive direction be *away* from the wall, and remember that velocity and momentum are vector quantities.

$$\text{Impulse} = F \Delta t = \text{change in momentum}$$
$$= 0.05 \, [10 - (-20)]$$
$$= 1.5 \, \text{N s}$$

Since impulse $= F \Delta t$, $1.5 = F \times 0.02$ giving $F = 75 \, \text{N}$.

Note: In impulse or change of momentum problems it is important to remember the vector nature of velocity and state the direction chosen as positive.

Experiment 1.5

The kicking of a football is often used to study impulse and change of momentum. A pupil has a strip of metal foil attached to the toe of his shoe and the ball has a similar strip at the position where it is to be kicked. Flexible wires connect the two strips to the terminals of an electric clock or scaler so that the time of contact Δt of the strips is recorded.

After impact the ball passes through a light beam operating another electric clock so that the time t for the ball's diameter D to pass through the beam may be measured.

The velocity v of the ball is then calculated from $v = \dfrac{D}{t}$. The mass m of the ball is measured on a balance. The force of impact F may then be calculated from

$$F = \frac{m \, \Delta v}{\Delta t}$$

where $\Delta v = v - 0$, the ball being initially at rest.

The force/time graph

A graph of force acting on an object against the time for which that force acts, can indicate the change of momentum of the object, or the impulse of the force.

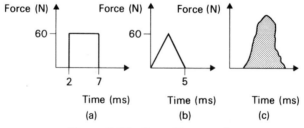

Figure 1.26 Force/time graphs

From Figure 1.26(a), the change of momentum, or impulse, is $F \Delta t = 0.3 \, \text{N s}$ or $0.3 \, \text{kg m s}^{-1}$ which is equal to the area under the graph.

From Figure 1.26(b) the force increases to a maximum, then decreases to zero giving an average force of 30 N and an impulse or a change of momentum of $0.15 \, \text{kg m s}^{-1}$, which is also the area under this graph. This applies in general for *any* shape of graph. The area under the graph of Figure 1.26(c) also gives the change of momentum or impulse.

Area under a force/time graph gives impulse or change of momentum

---

Newton's Third Law of Motion

If a body A exerts a force F on a body B, then B exerts a force of $-F$ on A. That is, action and reaction are equal and opposite. This pair of forces, on two bodies, both have the same magnitude but opposite directions. This law can be inferred from the conservation of linear momentum, see page 38.

Examples: The engine of a jet expels gas backwards and the jet moves forwards. The force on the jet in one direction equals the force on the gas in the opposite direction. A mass is attracted downwards by the Earth but the Earth is also attracted towards the mass by an equal force. However, as the Earth is large any infinitesimal movement caused will not be observed!

In questions involving *two* or more bodies and Newton's Third Law it is often necessary to apply $F = ma$ to *one* object, and consider *all* the forces on that object. Typical problems include tensions in cords, moving lifts and balances, and simple pulley systems.

Example: A 2 kg mass is suspended from a spring balance. If the reading on the balance is 30 N, what is the size and direction of the acceleration of the mass.

Consider the forces *on* the 2 kg mass. By Newton's Third Law, the downward pull on the spring balance of 30 N must equal the upward tension in the suspension cord of the 2 kg mass.

Figure 1.27

(1) Let upwards be positive

$$\text{Resultant force} = +10 \, \text{N upwards}$$
$$\text{Acceleration} = 5 \, \text{m s}^{-2} \, \text{upwards}$$

(2) Let downwards be positive

$$\text{Resultant force} = -10 \, \text{N downwards}$$
$$a = -5 \, \text{m s}^{-2} \, \text{downwards}$$

This answer implies a *deceleration* in the downwards direction.

In answer to the question: the size and direction of the acceleration is $5\,\mathrm{m\,s}^{-2}$ upwards. (But observe that a deceleration of $5\,\mathrm{m\,s}^{-2}$ downwards would also produce a reading of $30\,\mathrm{N}$ on the balance.)

Example: Two blocks of $2\,\mathrm{kg}$ and $1\,\mathrm{kg}$ are pulled by a $6\,\mathrm{N}$ force along a smooth horizontal surface. Calculate the acceleration of the system and the tension T in the connecting cord.

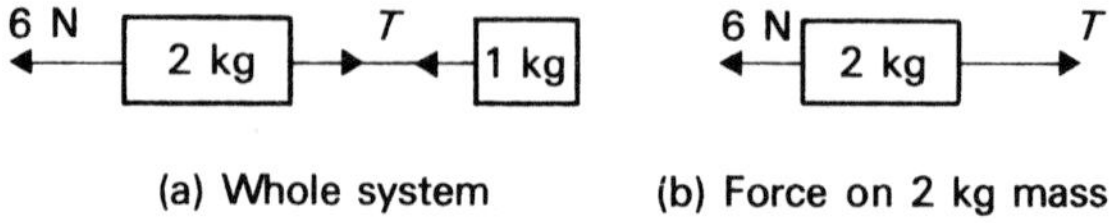

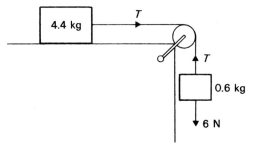

(a) Whole system (b) Force on 2 kg mass

Figure 1.28

For the *whole* system, using $F = ma$, $6 = 3 \times a$, giving $a = 2\,\mathrm{m\,s}^{-2}$.
Consider the $2\,\mathrm{kg}$ mass *alone* and *all* the forces acting on that mass.

$$6 - T = 2 \times 2 \qquad \text{using } F = ma$$
$$T = 2\,\mathrm{N}$$

There is a tension of $2\,\mathrm{N}$ in the connecting cord.

Example: A $4.4\,\mathrm{kg}$ block is attached to a $0.6\,\mathrm{kg}$ block by a light cord passing over a smooth pulley as shown in Figure 1.29. If the friction between the block and bench can be neglected, calculate the acceleration of the system and the tension in the cord.

T

4.4 kg

T

0.6 kg

$6\,\mathrm{N}$

Figure 1.29

Force on the system causing the movement $= 0.6 \times 10 = 6\,\mathrm{N}$.
Consider the *whole* system, and using $F = ma$

$$6 = 5 \times a \qquad \text{Total mass} = 4.4 + 0.6$$
$$a = 1.2\,\mathrm{m\,s}^{-2}$$

Consider the $4.4\,\mathrm{kg}$ mass *alone* and use $F = ma$

$$\text{Force } F = \text{tension}, \quad m = 4.4\,\mathrm{kg}, \quad a = 1.2\,\mathrm{m\,s}^{-2}$$
$$T = 4.4 \times 1.2 = 5.28\,\mathrm{N} \quad \text{towards the pulley}$$

or considering the $0.6\,\mathrm{kg}$ mass *alone*, the force downwards F is $6 - T$

$$\text{giving } 6 - T = 0.6 \times 1.2$$
$$T = 6 - 0.72 = 5.28\,\mathrm{N} \quad \text{towards the pulley}$$

The pull of the rope on the $4.4\,\mathrm{kg}$ mass is *equal* and *opposite* to the pull of the rope on the $0.6\,\mathrm{kg}$ mass.

Force as a vector

Often only part of a force is effective in causing an acceleration or a deceleration.

Example: A toy train, mass 0.2 kg, is given a push of 10 N at an angle of 30° to the rails. Find the effective forward force on the train. Hence calculate the acceleration of the train.

The 10 N force may be taken as composed of two forces at right angles, OA in a forward direction parallel to the rails, and OB at 90° to OA, tending to push the train over.

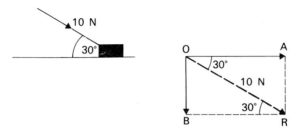

Figure 1.30

From the diagram the effective force is OA = 10 cos 30°, or it may be measured from the length of OA by accurate construction.

$$F = 10 \cos 30° = 10 \frac{\sqrt{3}}{2} = 8.66 \,\text{N}.$$

$$\text{Acceleration} = \frac{F}{m}$$

$$= \frac{8.66}{0.2}$$

$$= 43.3 \,\text{m s}^{-2}$$

Example: A small trolley, mass 300 g, is pulled up a plane inclined at 30° to the horizontal by a string parallel to the plane. What force is required just to prevent the trolley rolling backwards?

P—force up plane
W—weight of trolley due to gravity
Weight downwards $W = mg = 0.3 \times 10 = 3 \,\text{N}$.

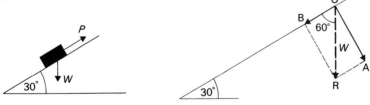

(a) Force diagram   (b) Component of weight OB down the plane

Figure 1.31

But the effective force, due to the weight, down the plane is required. Again the force OR can be split up into two forces at right angles (Figure 1.31(b)). OA, which is perpendicular to the plane, will have no effect along the plane. So OB is therefore the effective force down the plane.

$$OB = OR \cos 60°$$

Force down the plane $= OB = 3 \times 0.5 = 1.5\,N$.
A force of 1.5 N is required up the plane at P to prevent the trolley slipping backwards.

Note: In the above example the effective force, or *component of the weight*, down the plane OB, equals $W \times \cos R\hat{O}B$. Notice that for any plane.

$$\cos R\hat{O}B = \sin \text{ (angle of inclination of the plane } \alpha)$$

since $\alpha + R\hat{O}B = 90°$.
Because this angle α is the angle usually given in problems, it is useful to remember that the component of the weight down any plane of angle of inclination α to the *horizontal* is given by:

---

component of weight down the plane $= W \sin \alpha$

---

where W is the weight of the mass on the plane.

When a number of different forces are acting on an object the resultant force or vector sum of the forces can be determined by vector addition (see Section 1:A.3).

If an object is in equilibrium it will have no *unbalanced* force and therefore the vector sum of the forces must be *zero*.

Example: A picture of mass 2 kg hangs by two equal cords which make an angle of 60° as shown. What is the tension in each cord?

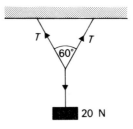

Figure 1.32

Since the picture is stationary the vector sum of the three forces must be zero. In a vertical direction the effective upward forces are $T \cos 30°$ for each cord.

Hence

$$2T \cos 30° = 20$$
$$T = 11.5\,N$$

1:C.3   *Work*

Unit: joule (J), scalar.
The work done is the effective force multiplied by the displacement over which

the force acts. Energy is transferred when a constant force moves its point of application. Energy is expended when an object is *moved*.

$$\text{Work} = \text{effective force} \times \text{displacement}$$
$$\quad\big|\qquad\qquad\big|\qquad\qquad\big|$$
$$\quad\text{J}\qquad\qquad\text{N}\qquad\qquad\text{m}$$

If an object moves in a certain direction and an oblique force is applied then only part of that force is effective in moving the object.

Example: The same toy train as in the example on page 29 is moved 2 m along its rails by a force of 10 N at 30° to the rails. Assuming no friction what is the work done?

$$\text{Effective force} = 10\cos 30°$$
$$= 8.66\,\text{N}$$
$$\text{Work done} = \text{force} \times \text{displacement}$$
$$= 8.66 \times 2$$
$$= 17.3\,\text{J}$$

Note: The definition of work involves displacement but in simple rectilinear problems the word distance may be encountered.

1:C.4 *Energy*

Unit: joule (J), scalar.
There are various forms of energy including: kinetic energy, potential energy, heat, sound, light, chemical, electrical, magnetic, and nuclear energy.

Kinetic energy, E_k

Kinetic energy is the energy an object has by virtue of its motion.

Derivation of $E_k = \frac{1}{2}mv^2$

An object in motion could do work in being brought to rest. Alternatively energy must be supplied, or work done, to give an object of mass m a velocity v. Consider the force required to give a velocity v after a displacement s.

E_k = work done, or energy required, to change the velocity from zero to v.

$$E_k = \text{force} \times \text{displacement} \qquad\qquad v^2 = u^2 + 2as$$
$$= ma \times s \qquad\qquad\qquad\qquad v^2 = 0 + 2as$$
$$= m\frac{v^2}{2} \qquad\qquad\qquad\qquad as = \frac{v^2}{2}$$

$$E_k = \tfrac{1}{2}mv^2$$
$$\text{J}\qquad\quad\text{kg}\quad\text{m}\,\text{s}^{-1}$$

Potential energy, E_p

This is the energy possessed by a body by virtue of its position or state of strain.

Derivation of $E_p = mgh$

An object requires work to be done to raise it to a higher position against the force of gravity.

E_p = force × displacement
= $mg × h$ (object is raised a vertical distance h)

$$E_p = mgh$$
$$J \qquad kg\, m\, s^{-2} \qquad m$$

   Potential energy is gained (or lost) when an object is raised (or falls) a *vertical* distance h, where the acceleration due to gravity is g. Thus the potential energy gained by an object raised a distance h on the Moon will be different to that gained on the Earth, as g does not have the same value on both.

Potential energy due to strain

Elastic objects, such as springs, may store potential energy. A spring may be extended or compressed and in these positions it has potential energy. When it is released this energy may be imparted to another object as kinetic energy, for example, the propulsion of vehicles by stretched cords on a linear air track.
   Experiment can show that the force F extending a spring, an elastic cord or a wire is directly proportional to the extension (ext) produced, providing the 'elastic limit' is not exceeded.

$$F \propto ext$$

   The **elastic limit** is that point beyond which an object does *not* return to its original shape and size when the deforming force is removed.
   The potential energy stored in a spring, extended a distance x, is equal to the work done in producing that extension.

E_p = work done
= average force × displacement
= $\frac{1}{2}$ final force × displacement
$E_p = \frac{1}{2}Fx$

The force is zero at the beginning and increases uniformly up to the final value F.

Note: A *graph* of force against extension for an elastic system will give a straight line through the origin. The potential energy stored in this system for an extension x will be the *area* under the graph for this value of x.

Potential energy E_p = area under force/extension graph

Experiment 1.6 *Varying acceleration*

A stretched elastic cord can be used to provide a changing force and hence a changing acceleration. A trolley is attached to a long extended elastic cord and a ticker tape as shown in Figure 1.33.

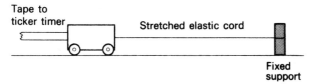

Figure 1.33

When the trolley is released, the force due to the stretched elastic will cause the trolley to accelerate towards the support. But this force will decrease with time, hence the acceleration will also decrease. Velocities are determined from the tape and a v/t graph is plotted which will *not* be a straight line. The shape of this v/t graph will be similar to the top left graph, acceleration decreasing, of Figure 1.14.

Another experiment which produces a non-uniform acceleration is illustrated by No. 22 in the Exercise section on page 284. Since the mass of the chain hanging vertically *changes*, the force on the system also varies.

Note: To produce a varying acceleration a changing force is required.

Heat

(1) The heat E_h transferred to a body of mass m when its temperature is raised by ΔT is given by:

$$E_h = mc\,\Delta T$$

where c is the specific heat capacity whose units are $\text{J kg}^{-1}\,\text{K}^1$.

(2) The heat E_h required to change the state of a body of mass m and specific latent heat l (whose units are J kg^{-1}) at the normal melting or boiling point is given as:

$$E_h = ml$$

Note: For definitions and discussions of specific heat capacity and specific latent heat (see Section 6:A.2).

Electrical energy

The electrical energy produced in a time t when a current I flows through a resistor R is given by:

$$E_{el} = IVt \quad \text{or} \quad E_{el} = I^2Rt \quad \text{(See Section 2:B.4)}$$

Conservation of energy

> The total energy of a closed system must be *conserved*
> although the energy may change its form.

An object held above the ground has potential energy but in falling it loses potential energy and gains kinetic energy until impact with the ground when it rebounds with a smaller kinetic energy (unless it is a perfectly elastic bounce). The energy lost is converted into heat, light, and sound.

Example: A ball of mass 0.2 kg is dropped from a height of 2 m above the ground and rebounds to a height of 1.8 m. What is the kinetic energy lost on impact?

$$E_k \text{ just before impact} = E_p \text{ at the start}$$
$$= 0.2 \times 10 \times 2$$
$$= 4 \text{ J}$$
$$E_k \text{ after impact} = E_p \text{ at top of first rebound}$$
$$= 0.2 \times 10 \times 1.8$$
$$= 3.6 \text{ J}$$
$$\text{Kinetic energy } E_k \text{ lost at impact} = 0.4 \text{ J}$$

Example: What velocity must a metal bullet have if it just melts when striking a target? (Assume that all the kinetic energy is turned into heat supplied to the bullet.)

Initial temperature of the bullet = 20°C
Melting point of metal = 320°C
Specific latent heat of fusion of the metal = $2 \times 10^4 \text{ J kg}^{-1}$
Specific heat capacity of metal = $120 \text{ J kg}^{-1} \text{ K}^{-1}$

$$E_k \text{ of bullet} = \tfrac{1}{2} mv^2 \qquad v\text{—velocity of bullet}$$

$$\text{Heat } E_h \text{ gained} = mc\,\Delta T + ml$$
$$= m \times 120 \times 300 + m \times 2 \times 10^4$$
$$= m(3.6 \times 10^4 + 2 \times 10^4)$$
$$= 5.6 \times 10^4 \, m \text{ J}$$

But all $E_k \longrightarrow E_h$

$$\tfrac{1}{2} mv^2 = 5.6 \times 10^4 \, m$$
$$v^2 = 11.2 \times 10^4$$

$$\text{Velocity of bullet} = 3.35 \times 10^2 \text{ m s}^{-1}$$

Simple problems involving the conversion of mechanical and heat energy are important and should be understood. Notice that the actual value of the mass is not always required.

Experiment 1.7 *Interchange between potential energy and kinetic energy*

When an object slides down a smooth plane, or a pendulum swings down from a height h, as shown in Figure 1.34, the potential energy is converted into kinetic energy. If there is no energy loss due to friction or air resistance then:

$$E_p \longrightarrow E_k \quad \text{giving } mgh = \tfrac{1}{2}mv^2$$
$$\text{and } h \propto v^2$$

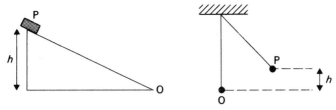

Figure 1.34 Potential energy into kinetic energy

where v is the velocity at point 0, a *vertical* distance h below the starting point P, the velocity at P being zero.

The value of h can be measured directly and the velocity v determined using a photocell/scaler method. A graph of h against v^2 will give a straight line which will pass through the origin if the friction is negligible. The pendulum arrangement will yield the better result since friction between the object and plane is *unlikely* to be negligible. (See Problem 1.23 page 43).

Similarly when a vehicle or object is propelled by elastic cords, the potential energy of the elastic is converted into kinetic energy of the vehicle. Since each elastic cord can represent one unit of potential energy, the total number of cords N varies as the total E_p, hence if all $E_p \longrightarrow E_k$ then:

$$N \propto v^2$$

where N is the number of cords which impart a velocity v to the vehicle. Again, the velocity is determined by a photocell/scaler method and a graph of N and v^2 should give a straight line through the origin.

Energy changes in oscillatory systems

A pendulum, such as that represented in Figure 1.35, executes a regular to-and-fro motion between positions A and B because the force on the bob is always directed towards O. The resulting acceleration causes the bob to attain a velocity at O which carries it through to the other side.

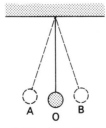

Figure 1.35 Simple pendulum

The force and the acceleration are both zero at O, but the velocity and the kinetic energy E_k are both at their maximum values. At A or B the velocity, and therefore the kinetic energy E_k have become zero, this energy having been converted to potential energy E_p which is then at its maximum value. The pendulum is thus a device in which there is a continual transfer of energy between the kinetic and potential forms.

> This regular motion is called an **oscillation** or a **vibration** and acts about a mean position O, with a continual interchange of potential and kinetic energy.

A **sonometer** wire may vibrate in various modes, depending where it is plucked or stimulated. The resulting frequencies are called the **fundamental** and **overtones**. Higher overtones may be obtained emitting correspondingly higher frequencies $4f_0$, $5f_0$, ...

Fundamental frequency f_0 1st overtone $2f_0$ 2nd overtone $3f_0$

Figure 1.36   Modes of vibration of a wire fixed at both ends

Experiment 1.8   *To investigate the frequency of oscillatory systems*

A number of vibrating systems are set up, as shown in Figure 1.37, and their frequencies determined as the following physical properties are altered.

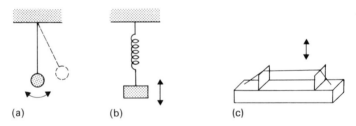

(a)   (b)   (c)

Figure 1.37   Vibrating systems

In (a) the length of the pendulum.
In (b) the mass of an object vibrating vertically on the end of a spring.
In (c) the tension of a sonometer wire.

The results of these experiments show first, that each system has a **natural** or **fundamental frequency** and secondly, that this frequency is only *changed* by *altering* the *physical properties* of that system. When any system is set into motion it will vibrate with its natural frequency, or possibly with a higher multiple called an overtone, as with the sonometer wire, but the amplitude of the natural frequency will be the greatest.

Damping

A system will continue to vibrate with the same amplitude, that is the displacement from the rest or mean position, providing no **energy** is lost from the system. If the amplitude decreases the oscillation is said to be **damped.** For example, air resistance will eventually cause the pendulum to come to rest. Most natural systems exhibit damping.

The *energy losses* causing damping may be due to:
(1) internal or external friction: for example, energy lost internally through heating, or external air or water resistance.
(2) radiation: for example, a plucked string may emit a sound wave, or an oscillatory circuit may emit electromagnetic radiation.

To overcome these losses and *maintain* the oscillations, **energy** must be fed in, but at an appropriate frequency. If a pendulum were oscillating at 20 Hz an impulse could be given 20 times a second to coincide with its natural swing. Impulses given at the rate of 11 times a second would interfere with its natural vibration and cause damping.

Note: To maintain the oscillations and overcome damping, energy must be fed in at the *same* frequency as the natural frequency of the system.

Resonance

Resonance is the response of a system, when the forcing frequency is equal to the natural frequency and results in an increase in the amplitude of the oscillation. The frequency at which this occurs is termed the **resonance frequency**. The resonance frequency is *equal* to the natural frequency of the system.

1:C.5 *Power and Efficiency*

Power

Unit: watt (W), scalar.
Power is the rate of doing work, or the rate of expending energy.

$$\underset{\underset{\text{W}}{|}}{\text{Power}} = \underset{\underset{\text{s}}{|}}{\overset{\overset{\text{J}}{|}}{\frac{\text{work done}}{\text{time taken}}}} \quad \text{or} \quad \overset{\overset{\text{J}}{|}}{\underset{\underset{\text{s}}{|}}{\frac{\text{energy expended}}{\text{time}}}}$$

$$\text{Power} = \frac{\text{force} \times \text{displacement}}{\text{time}}$$

$$\text{Power} = \text{force} \times \text{velocity}$$

Attention must still be given to the direction of the force. Only the effective force in the direction of the velocity must be used in calculating the power.

Efficiency of a machine

$$\text{Efficiency} = \frac{\text{energy out}}{\text{energy in}} \times 100\%$$

This is a percentage and has no units.

Example: A 0.6 kW motor is used to raise a 50 kg block up a vertical height of 8 m. How long does it take? State any assumptions made.

$$\text{Power} = \frac{50 \times 10 \times 8}{\text{time}}$$

$$0.6 \times 10^3 = \frac{4 \times 10^3}{t}$$

$$\text{Time} = 6.7 \text{ s}$$

100 % efficiency has been assumed.

Example: A train of mass 10^5 kg accelerates from rest to 40 m s^{-1} in 20 s. The total frictional resistance is 0.5×10^5 N. Calculate (a) the total force provided by the motor and (b) the power developed at the end of the sixteenth second.

(a) Acceleration $= \dfrac{40}{20} = 2 \text{ m s}^{-2}$

Force required to produce acceleration $= 10^5 \times 2 = 2 \times 10^5$ N
Total force provided by the motor $= 2 \times 10^5 + 0.5 \times 10^5 = 2.5 \times 10^5$ N

(b) Power $=$ force $\times$ velocity
$= 2.5 \times 10^5 \times 32$
$= 8 \text{ MW}$

(Velocity at the end of the sixteenth second $= 32 \text{ m s}^{-1}$.)

1:C.6 *Momentum, Conservation of Momentum, and Collisions*

In a collision the time interval during which the objects are in contact is small.

Conservation of momentum

(For a definition of momentum see Section 1:C.2 page 24.)

> Provided that no external force acts the total momentum
> in *any* collision is conserved.

This law is consistent with Newton's Third Law. For an object A colliding with an object B producing changes in velocity Δv_A and Δv_B respectively,

$$F_A = -F_B$$

$$\frac{m_A \, \Delta v_A}{\Delta t} = -\frac{m_B \, \Delta v_B}{\Delta t}$$

Δt—time interval over which the force acts during the collision

Thus $m_A \, \Delta v_A + m_B \, \Delta v_B = 0$

Therefore the *change* of momentum of A + the *change* of momentum of B is zero. Hence there is *no* gain or loss of momentum in the collision.

Types of collision

(1) **Elastic collision:** Momentum *and* kinetic energy are conserved. For

example, a ball bouncing off the ground back up to its original height; a moving nucleus deflected by another nucleus.

(2) **Inelastic collision:** Momentum is conserved but the kinetic energy usually *decreases*, being converted into heat or elastic potential energy, causing deformation for example, a ball dropped on to sand or mud. In a *completely* inelastic collision, the two objects join together after impact for example, a vehicle colliding with and joining on to another vehicle.

(3) **Explosions:** Momentum is conserved but the kinetic energy *increases*. For example, when two trolleys are made to fly apart the potential energy stored in the spring is converted into the kinetic energy of the trolleys. (Remember that momentum is a vector quantity. It will be zero before and after the explosion only when account is taken of the direction of the velocities.)

Note: Momentum is conserved in all types of collisions but kinetic energy is only conserved in elastic collisions.

Example: A red ball, mass 30 g, travelling at $4 \, \text{m s}^{-1}$, collides with a stationary green ball, mass 50 g, which moves off with a velocity of $2 \, \text{m s}^{-1}$ in the same direction as the red ball. What is the final velocity of the red ball? Is the collision elastic?

Momentum is conserved.

$$\text{Momentum before collision} = \text{momentum after collision}$$
$$30 \times 4 + 50 \times 0 = 30 \times v + 50 \times 2$$
$$v = \tfrac{2}{3}$$
$$= 0.67 \, \text{m s}^{-1}$$

Notice that the masses have *all* been left in grams.

Using $E_k = \tfrac{1}{2}mv^2$, but with the masses now in kg,

$$E_k \text{ before collision} = \tfrac{1}{2} \times 0.03 \times 4^2$$
$$= 0.240 \, \text{J}$$
$$E_k \text{ after collision} = \tfrac{1}{2} \times 0.03 \times (\tfrac{2}{3})^2 + \tfrac{1}{2} \times 0.05 \times 2^2$$
$$= 0.107 \, \text{J}$$

The collision is *not* elastic because there has been a change in kinetic energy.

Note: Momentum is a *vector* quantity hence it is important to state the positive direction and remember that velocities in the reverse direction are negative.

Example: A 0.6 kg trolley travelling at $2 \, \text{m s}^{-1}$ collides with a 0.7 kg trolley moving at $1 \, \text{m s}^{-1}$ in the opposite direction. After collision the two trolleys lock together. Determine their velocity.

Let the initial direction of the 0.6 kg trolley be positive.

By the conservation of momentum:

$$(0.6 \times 2) - (0.7 \times 1) = (0.6 + 0.7)v$$
$$0.5 = 1.3v$$
$$v = 0.38$$

Since the calculated value of v is positive, the final velocity is $0.38 \, \text{m s}^{-1}$ in the original direction of the 0.6 kg trolley.

Experiment 1.9

Using vehicles on a linear air track (to avoid friction) the velocities before and after elastic and inelastic collisions may be determined experimentally using a scaler connected to suitably placed photodiodes. The conservation of momentum can then be verified for elastic and inelastic collisions. The kinetic energy may be calculated for each collision and its conservation in elastic collisions only may be verified.

Experiment 1.10

To study an explosion, two trolleys with ticker tapes attached may be made to fly apart. Their velocities after the explosion may then be determined and the vector sum of the final momenta may be shown to equal zero.

Momentum as a vector oblique collisions*

Since momentum is a vector quantity it follows that if the colliding objects are not in the same straight line the total momentum before and after a collision must be found from the *vector* addition of the separate momenta.

Example: An object A collides with a stationary object B at O. Initial momentum = $m_a u_a$. The final momentum is obtained by the vector addition of $m_a v_a$ and $m_b v_b$, as shown in Figure 1.38. Hence the length of PR will correspond to the total final momentum. As momentum is conserved the final momentum must equal the initial momentum in magnitude and direction and so the direction of PR will be the same as that of $m_a u_a$, the initial momentum of A.

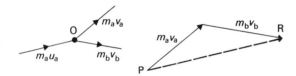

Figure 1.38

The same principles are applied in nuclear collisions, for example, when an α particle collides with a helium nucleus. The velocities of the particles may be determined from cloud chamber photographs. These collisions may usually be considered as elastic, if no change in internal structure occurs, and if the particles are not captured.

Figure 1.39 A cloud chamber photograph showing the paths of α particles in a chamber containing helium. Notice the angle between the directions of motion after the collision. (*P.M.S. Blackett, F.R.S., Proc. Roy. Soc.,* 107A, **349**, 1925)

---

* Oblique collisions are not required for examination purposes.

Note: If an object A collides obliquely with a stationary object B of *equal mass* and the collision is *elastic*, the angle between the directions of motion of A and B after the collision will be 90°. If *either* the masses are not equal *or* the collision is inelastic the angle may *not* be 90°.

Conservation laws

Certain physical quantities are found to be **conserved**, that is, the total value of the quantity is the same at the beginning as at the end of an experiment.

Example: If a metal object of charge $+3$ units is touched by another metal object carrying a charge of -8 units, the final charge is observed to be -5 units. The *total* charge has not altered.

Also, if a cloth is rubbed against a plastic rod, the cloth and the rod can become charged. A *separation* of positive and negative charge is achieved, but no charge is created.

A *closed system* is one in which all the constituent objects or apparatus involved are included. For example, in a specific heat capacity experiment, all the apparatus, liquid, container, thermometers *and* the surrounding air, must be considered when discussing heat transfer. In a chemical reaction, the masses of all the reacting substances must be included whether solid, liquid, or gaseous.

Three conservation laws are listed below. The conservation of mass for non-nuclear chemical reactions is a special case of the conservation of energy, because of the equivalence of mass and energy, (see Section 7:C.1).
(1) **Conservation of energy**
    The total energy of any closed system is conserved.
(2) **Conservation of linear momentum**
    Providing no external forces act, the total momentum in any collision is conserved.
(3) **Conservation of charge**
    The total charge of any closed system is conserved.
Only the specific case of the conservation of linear momentum in collisions is given here. Rotational motion, e.g. objects spinning on an axis, or moving around a circle, are not discussed at this level. For interest, it may be mentioned that there is conservation of *angular* momentum.

Example: A ball rolls from rest down a slope of vertical height h. Discuss the conversion of the initial potential energy into other forms.

Energy is conserved. The potential energy at the top is converted into kinetic energy at the bottom, together with some heat due to the friction between the ball and the slope. It should be mentioned that the kinetic energy will be partly *translational* E_k(trans) and partly *rotational* E_k(rot).

$$E_p(mgh) \longrightarrow E_k(\text{trans}) + E_k(\text{rot}) + \text{heat}$$

At this level in problem solving, the rotational kinetic energy can be ignored.

Problems

1.13 A 2 kg mass is suspended by a balance from the roof of a lift. The scale of the balance is from 0 to 40 N. What will the balance read:
(a) when the lift is at rest,
(b) when the lift is moving upwards at a constant velocity of $3\,\text{m s}^{-1}$,
(c) when the lift is accelerating upwards at $2.5\,\text{m s}^{-2}$?

1.14   A boy of mass 50 kg stands on scales placed on the floor of a lift. The lift starts from rest and then the scales read 750 N. Describe the motion of the lift.

1.15   Two objects are joined by a rope passing over a smooth pulley. The 2 kg mass is held in the position shown in Figure 1.40.

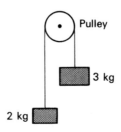

Figure 1.40

(a) Determine the acceleration of the system when the 2 kg mass is released.
(b) Determine the tension in the rope.
Hint: make two sketches showing the forces on each mass separately.

1.16   A catapult has a horizontal force of 320 N on a steel ball of mass 0.02 kg when extended back a horizontal distance of 10 cm.
(a) What is the initial acceleration of the ball?
(b) Calculate its average acceleration when in contact with the catapult.
(c) Calculate the velocity of the ball when leaving the catapult.
(d) What is the effect on this velocity if two identical cords were used, in the same arrangement, doubling the force? State any assumptions made.

1.17   A ball of mass 2 kg moves in a straight line with a velocity of $20 \, \mathrm{m\,s^{-1}}$. A constant force acts for 3 s changing the velocity to $4 \, \mathrm{m\,s^{-1}}$ in the opposite direction. Calculate:
(a) the initial momentum of the ball,
(b) the impulse acting on the ball,
(c) the magnitude and direction of the force, and
(d) the displacement of the ball.

1.18   A trolley of mass 2 kg is released from rest 6 m from the bottom of a slope which makes an angle of 30° to the horizontal. The constant frictional force opposing the trolley is 4 N.
(a) What is the resultant force on the trolley down the plane?
(b) Calculate the speed of the trolley at the bottom of the plane.
(c) If the trolley then moves along a horizontal table, at the bottom of the plane, which has the same frictional force, how far along the table will the trolley travel before coming to rest?
(d) Draw a graph of speed against time for the trolley's journey from the top of the plane until it comes to rest. Numerical values are required on the axes.

1.19   Forces A, B, and C of magnitude 6 N, 9 N, and 12 N respectively act from a point O and are in equilibrium. Find the size of angle AÔB.

1.20 g is the acceleration due to gravity on Earth. A 3 kg block is released from rest on a planet where the gravitational field strength is $\frac{g}{4} \, \mathrm{N\,kg^{-1}}$. What is the velocity of the block after 6 s? Calculate the decrease in potential energy of the block after these 6 s.

1.21   A small pellet of mass 0.47 g is fired into a piece of plasticine attached to a vehicle on a linear air track. After the pellet hits the plasticine the vehicle moves forward and a card of length 12 cm, joined to the vehicle, interrupts a light beam directed onto a photocell. The photocell is connected to the STOP terminals of a scaler.

Given: mass of vehicle plus plasticine and pellet = 200 g
time of interruption of light beam = 0.25 s
mass of gun = 1.1 kg

(*a*) Calculate the velocity of the bullet after being fired.
(*b*) Calculate the recoil velocity of the gun.

1.22

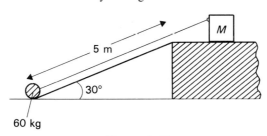

Figure 1.41

A 60 kg block is raised onto a platform by a machine M. The frictional force between the block and the slope is 11 N. If the machine is 80% efficient, calculate the power rating of the machine M if the operation takes 3 minutes.

1.23 In an experiment to study the conversion of potential energy to kinetic energy, a 2 kg box, initially at rest, was allowed to slide down a plank, one end of which was raised a vertical distance *h*. The velocity of the box at the bottom of the plank was determined. The following results were obtained.

distance *h*	0.5 m	1.5 m
velocity *v*	$2.0\,\mathrm{m\,s^{-1}}$	$4.9\,\mathrm{m\,s^{-1}}$

(*a*) Determine the potential energy at the top of the slope and the kinetic energy at the bottom for each pair of results.
(*b*) From these calculations estimate the energy 'lost' due to friction and state what happened to this 'lost' energy.
(*c*) Calculate the result likely for *v* when *h* = 1 m.

1.24 A stationary hard ball X of mass 0.1 kg is struck by a cricket bat and moves off with a speed of *v*. Another stationary ball Y of the same mass is struck and moves off

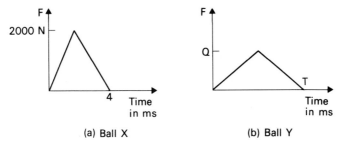

(a) Ball X   (b) Ball Y

Figure 1.42

with the same speed. Graphs of the variation of the force on each ball with time are given in Figure 1.42.

(a) State any differences in the balls X and Y.

(b) What physical quantity does the area under the graph represent?

(c) Calculate the speed at which the balls leave the bat.

(d) Suggest values for Q and T, state clearly the reasons for your choice.

1.25 State the physical quantities which correspond to the area under the following graphs:

(a) velocity/time,

(b) acceleration/time,

(c) force/time,

(d) force/displacement.

2 Electricity

A: Electric Charge, the Electric Field, and Movement of Charge in an Electric Field

2:A.1 Electric Charge

Unit: coulomb (C), scalar.
The charge on an electron is the smallest quantity of charge observed and is equal to 1.602×10^{-19} coulomb.

Note: The ampere is the basic electrical unit, hence the coulomb is defined as that charge which passes a point in a conductor in one second when a current of one ampere is flowing.

A neutral atom consists of a central positive nucleus and a number of electrons whose total charge is numerically equal to that of the nucleus. If some of these electrons are removed a positive ion is formed, whilst when there is an excess of electrons a negative ion is obtained. A positively charged body attracts a negatively charged one and like-charged objects repel each other.

For any two objects with charges Q_1 and Q_2 there will be a force F of attraction or repulsion depending on the *magnitude* of the charges, $F \propto Q_1 Q_2$. Also, this force will depend on the *separation d* of the charges. If the charges are only 1 cm apart the force will be greater than if the separation is 1 m. Experiment can confirm that the force varies *inversely* as the *square* of the separation, $F \propto \dfrac{1}{d^2}$.

2:A.2 The Electric Field

Charged objects will produce an **electric field**, an *electrostatic* field, in the surrounding region. Any other charge placed in that region will experience a force of attraction or repulsion. The *magnitude* of the force on a 'test' charge is an indication of the electric field at that point. The *direction* of this force, that is the direction of the field, will depend on the distribution of the positive and negative charges responsible for the field. The idea of a field was introduced in connection with the gravitational field.

Electric field strength

Units: $\dfrac{\text{newton coulomb}^{-1}}{\text{volt metre}^{-1}}$, vector.

The electric field strength E at any point is the force on a unit positive charge placed at that point.

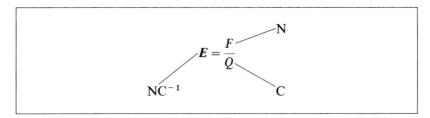

Students are reminded not to confuse electric field strength E, a vector quantity, with energy E which is scalar.

By convention, the *direction* of the electric field is the direction in which a free *positive* charge would move. A *line* of electric field shows the direction of the field. A free positive charge will move away from another positive charge or

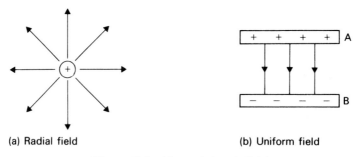

(a) Radial field   (b) Uniform field

Figure 2.1   Lines of electric field

towards a negative charge or surface. Figure 2.1(a) shows a **radial** field with lines of electric field symmetrically radiating outwards. Figure 2.1(b) shows a **uniform** field with lines of electric field parallel to each other.

Potential difference

Unit: volt, scalar.
The potential difference (p.d.) *between* two points is the work done, or energy expended, in transferring unit charge between those points.

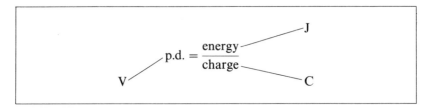

In any electric field, that is in a region containing charge, some work must be done or energy be expended, if charges are moved against the field. In Figure

2.1(b) a negative charge must be pushed from plate A to B. The gravitational field provides an analogy: the top of a hill has a higher gravitational potential than the bottom and energy must be expended in taking a mass up the hill. Conversely, potential energy is lost when the mass falls down the hill. It is the *difference* in the heights which determines the amount of energy required, the actual heights above sea level or the base of the hill are unimportant. Similarly in the electrical case the potential difference is between two points in a circuit, the p.d. or voltage is *not* at a point or through a wire.

Note: The potential difference is a scalar quantity as it has no direction.

Electrical energy = potential difference between two points
× charge moved between those points.

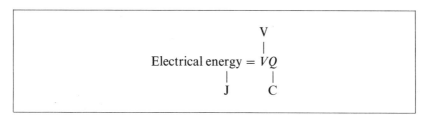

Relationship between potential difference and electric field strength for a uniform field only

In a *uniform* field the value of the electric field strength remains constant at *all* points between the plates. The lines of electric field are parallel to each other, therefore any charge moving between the plates A and B in Figure 2.2 will travel parallel to the lines of electric field. There are no 'sideways' forces to consider.

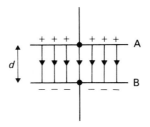

Figure 2.2   Uniform electric field

Potential difference between plates A and B a distance d apart

V = work done to take a unit positive charge from B to A

$$V = \frac{\text{work done B} \rightarrow \text{A}}{\text{charge}}$$

$$V = \frac{\text{force} \times \text{distance AB}}{\text{charge}}$$

V = electric field strength × distance AB

$$V = E \times d$$

$$\underset{\text{V}}{\diagup} \quad \underset{\text{V m}^{-1}}{\diagup} \quad \underset{\text{m}}{\diagdown} \qquad \text{or} \quad E = \frac{V}{d}$$

The quantity $\dfrac{V}{d}$ is called the **potential gradient**.

This equation shows how the other unit of electric field strength, volt metre^{-1}, is obtained.

Note: The electric field is a vector quantity, therefore a different treatment is required if the direction of the field is not parallel to the perpendicular distance d between the plates.

2:A.3 *Movement of Charges in an Electrical Field*

Only movement of charges in a *uniform* electric field will be considered.

Note: In a uniform field the electric field strength is constant. This implies that the *force* on a given charge is the *same* at any point in that field. Thus the force on a given charge near A, near B, midway or at any point between A and B in Figure 2.3 is exactly the same.

Charges moving parallel to an electric field

Since an electron has a negative charge it will freely travel from B to A in Figure 2.3. The acceleration gained by the electron will depend on the magnitude of the field.

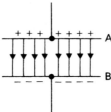

Figure 2.3   Electron accelerates towards A

For problems involving force and acceleration, the definition of electric field strength is used.

Example: An electron is at plate B, in Figure 2.3. What will be its acceleration towards A if the field is 2000 V m^{-1}, given that the charge on an electron is 1.6×10^{-19} C and its mass is 9×10^{-31} kg?

$$\text{Force on the electron} = E \times \text{charge on the electron}$$
$$= 2000 \times 1.6 \times 10^{-19}$$
$$= 3.2 \times 10^{-16} \text{ N}$$

From Newton's Second Law, $F = ma$

$$a = \frac{3.2 \times 10^{-16}}{9 \times 10^{-31}}$$

$$= 3.5 \times 10^{14} \text{ m s}^{-2}$$

Although the force is small, the acceleration is large because of the extremely small mass of the electron. The downward acceleration due to gravity may be neglected here in comparison!

Energy considerations

An electron travelling from B to A in Figure 2.3 will lose electrical energy and gain kinetic energy. By considering this interchange of energy the velocity can be calculated.

Example: An electron leaves an electron gun with negligible velocity. The potential difference between the gun and the collector plate is 100 V. What is its velocity just before it touches the collector plate?

$$\text{Electrical energy} = VQ$$
$$= 100 \times 1.6 \times 10^{-19}$$
$$= 1.6 \times 10^{-17} \text{ J}$$
$$\text{Kinetic energy gained} = \tfrac{1}{2}mv^2$$
$$= \tfrac{1}{2} 9 \times 10^{-31} v^2 \text{ J}$$

Assuming all the electrical energy is converted to kinetic energy

$$\tfrac{1}{2} 9 \times 10^{-31} v^2 = 1.6 \times 10^{-17}$$

Hence
$$v^2 = 0.35 \times 10^{14}$$
and
$$v \simeq 6 \times 10^6 \text{ m s}^{-1}$$

In general, for calculations involving forces and accelerations, consider the electric field, when velocities, times, and distances are required, it is more useful to consider the energy and potential difference. With high electric fields and light particles the acceleration due to gravity may be negligible but with heavier, slower-moving particles it must be included in the calculation.

Charges moving initially perpendicular to an electric field

For a projectile, the motions in any two perpendicular directions can be treated independently. When an electron enters an electric field it will experience a force attracting it to the positive plate, until it leaves the field.

Example: An electron travelling at 2×10^6 m s^{-1} enters a uniform electric field of 400 N C^{-1}. The electron is moving initially perpendicular to the field, see Figure 2.4(a). The length of plates A and B are 3 cm.
Determine:
(a) the perpendicular component of the velocity on leaving the field, (i.e. the component of the velocity in a direction parallel to the plates A and B).
(b) the time the electron spent in the field.
(c) The component of the velocity parallel to the field after passing between the plates.
(d) the new direction of motion.

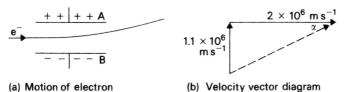

(a) Motion of electron   (b) Velocity vector diagram

Figure 2.4

(a) The perpendicular component remains at $2 \times 10^6 \, \mathrm{m\,s^{-1}}$.

(b) The perpendicular component of the velocity remains constant hence the time to travel $0.03 \, \mathrm{m}$ is $t = 1.5 \times 10^{-8} \, \mathrm{s}$.

(c) Parallel to the field there is a force F

$F = E \times Q = 400 \times 1.6 \times 10^{-19} = 6.4 \times 10^{-17} \, \mathrm{N}$
acceleration $= 6.4 \times 10^{-17}/9 \times 10^{-31} = 7.1 \times 10^{13} \, \mathrm{m\,s^{-2}}$
$u = 0$ (parallel to the field), $t = 1.5 \times 10^{-8}$
using $v = u + at$ $\qquad v = 1.1 \times 10^6 \, \mathrm{m\,s^{-1}}$

(d) The new direction of travel is obtained by drawing a vector diagram, Figure 2.4(b), and determining the angle α.

$$\tan \alpha = 1.1 \times 10^6 / 2 \times 10^6 \quad \text{giving } \alpha = 29°$$

Remember that when the electron *leaves* the field the force is *zero* and the velocity will remain unchanged. The acceleration due to gravity can be neglected compared to $7.1 \times 10^{13} \, \mathrm{m\,s^{-2}}$.

Comparison of motion in an electric field and a gravitational field

When a mass is projected horizontally near the Earth's surface, its velocity and position can be determined by considering the horizontal and vertical components separately, as described on page 18. Neglecting air resistance the horizontal component remains constant, but vertically the mass will accelerate under gravity, resulting in a *curved* trajectory towards the ground.

For a charged particle projected perpendicular to an electric field, a similar *curved* trajectory is obtained between the plates. There is acceleration due to the electric field in the 'parallel direction' but constant velocity in the 'perpendicular direction'. However, once the charged particle leaves the field its velocity will not alter. In the previous example, the electron will travel in a *straight line* at an angle of 29° to the plates.

Note: Compare the magnitude of the gravitational and electrostatic forces. A gravitational force ($F = mg$) can be hundreds of newtons, but the electrostatic force ($F = EQ$) on an electron is very small.

The discrete nature of charge

Any charge Q is an integral multiple of the electronic charge. This discrete nature of charge is used in Millikan's oil drop experiment to determine the *value* of the electronic charge.

Experiment 2.1 *Millikan's oil drop experiment*

Small drops of oil, released between the plates, are charged by friction as they emerge from the nozzle (Figure 2.5). One drop is singled out and viewed by the microscope. The

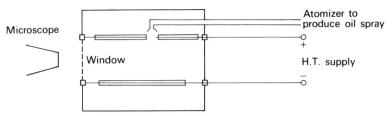

Figure 2.5 Apparatus for Millikan's oil drop experiment

drop will fall under gravity but the air resistance will provide an upward viscous force F_v and the drop will attain a steady terminal velocity, v.

Stoke's Law gives a relationship between the viscous force F_v and the terminal velocity v of a drop of radius r, namely

$$F_v = 6\pi\eta rv$$

where η is the viscosity of the medium.*

At the terminal velocity, but neglecting Archimedean upthrust

$$\text{viscous force} = \text{weight of drop}$$
$$F_v = mg$$
$$F_v = 6\pi r\eta v = mg$$
$$6\pi r\eta v = \tfrac{4}{3}\pi r^3 \rho$$

where r = radius of drop, ρ = density of oil, m = mass, and η = viscosity of air. Measurement of this terminal velocity together with accepted values of η, and ρ enables the radius of the drop, and hence the mass, to be determined.

A potential difference V_1 is now applied between the plates and adjusted so that the drop is brought to rest. The charge on this *same* drop is then altered by using a radioactive source or X-rays and the new potential difference V_2 required to bring the drop to rest is again recorded. This is repeated and the potential differences recorded.

Upward force due to the electric field $= \dfrac{V}{d}Q$ where Q is the total charge on the drop

and d the distance between the plates.

At rest, the weight $mg = \dfrac{V}{d}Q$ hence for each reading:

$$\frac{mgd}{V_1} = Q_1 = n_1 e$$

where $n_1, n_2, \ldots$ are integers as the total charge on the drop is an integral number of electronic charges. The values of $Q_1, Q_2 \ldots$ can be calculated for each $V_1, V_2 \ldots$ since m, g and d can be determined. By inspection of the Q values the value of e may be ascertained. As a charged drop will tend to repel the addition of further charges the *number* of electrons $n_1, n_2, \ldots$ on the drop will tend to be small.

Note:

(1) A Millikan type experiment demonstrates the *particulate* nature of electric charge. The total charge is always an integral multiple of a minimum quantity of charge, namely the electronic charge e⁻; i.e. the smallest 'particle' of charge is this electronic charge.

(2) Because of this particulate nature of charge the *value* of the electronic charge in coulombs can be determined by this experiment.

In the original experiment Millikan applied an electric field and measured the new terminal velocity of the drop, v_1. He then altered the charge on the drop and

---

*Memorization of the formula for viscosity is not required, but it is given to show that v must be measured to obtain the radius r.

in each case measured the new terminal velocity $v_2, v_3, \ldots$ He did *not* alter the potential difference and bring the drop to rest as described in the laboratory experiment above.

Problems

2.1 An electron, of mass 9×10^{-31} kg and charge 1.6×10^{-19} C, in a cathode ray tube, is accelerated by a uniform field of 3000 V m^{-1} between the anode and cathode which are 5 cm apart. It continues to the screen 10 cm beyond the anode.

(a) With what velocity does an electron reach the anode if it leaves the cathode with zero velocity and its mass remains constant?

(b) Assuming there is negligible electric field beyond the anode, what is the velocity of the electron at the screen?

(c) Calculate the time taken to travel from cathode to anode, and anode to screen.

(d) How would the velocity at the anode and the transit time from cathode to anode be affected if the distance between the electrodes was kept constant but the electric field were increased?

2.2 How many electrons are there in one coulomb?

2.3 (a) State why the terminal velocity of the drop in a Millikan-type experiment to determine the electronic charge, must be measured before the electric field is applied.

(b) In such an experiment the following results for charges on the drop were obtained, in 10^{-19} C: 3.18, 12.80, 4.79, 6.41, 8.01. What is a likely value for the electronic charge?

2.4 A uniform electric field is maintained by a p.d. of 20 V applied across two parallel horizontal plates of length 10 cm placed 4 cm apart in a vacuum. A proton travelling at 10^5 m s^{-1} parallel to the 10 cm length plates, enters the field. After leaving the field, the proton continues to a collector 40 cm beyond the plates. (Mass of proton $= 1.67 \times 10^{-27}$ kg).

(a) Draw and label a sketch to illustrate the direction of travel of the proton over these 50 cm and describe the motion over each section.

(b) Calculate the horizontal and vertical components of the velocity on leaving the electric field.

(c) Determine the velocity of the proton at the collector.

2.5 Two plates 0.1 cm apart have a potential difference of 100 V applied between them. A small oil drop of mass 1.44×10^{-11} g is held midway between the plates.

(a) Draw a sketch to indicate all the forces acting on this drop.

(b) State the magnitude of each force, giving a reason where necessary.

(c) State the number of electrons on the drop.

B: Current, Resistance, Electromotive Force, Electrical Energy, and Power

2:B.1 *Current*

In some substances, for example copper, the outer electrons of the atom are less strongly bound to the individual nuclei and may wander throughout the bulk of the material. Such a substance is termed a **conductor**, see Figure 2.6. If a potential difference is applied between two points on such a substance these conduction electrons will travel to the positive point, that is the point of highest potential giving a transient current (i.e. a current of short duration). If this potential difference is maintained then a steady current is produced.

In terms of *electric field strength*, which depends on the p.d. between A and B as well as the separation of A and B, there is a force on any mobile charges, causing them to move. In this case the electrons experience a force and move towards A.

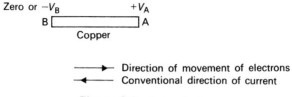

Figure 2.6   A conductor

Note: By convention the direction of the current is the direction in which *positive* charge would move. Hence the current flows from a point of high potential A to a point of lower potential B, but electrons, the carriers of charge in this instance, move in the opposite direction.

If a substance does not have these conduction electrons then a moderate potential difference applied across it will not produce a current and the substance is termed non-conducting, or an **insulator**.

Current

Unit: ampere (A), scalar.
The current flowing in a conductor is the total charge passing a given cross section in one second and is dependent on the *number* of charge carriers together with their speed.

$$I = \frac{Q}{t}$$

The unit of current, the **ampere**, is the basic electrical unit and is defined in terms of the force produced between two current carrying conductors. It is observed that when two current-carrying wires are placed adjacent to each other there is a force of attraction between them, if their currents are in the same direction (Figure 2.7(a)). When the currents are in opposing directions the force is one of repulsion (Figure 2.7(b)).

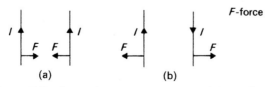

Figure 2.7   Forces between current carrying conductors

The ampere is defined as that current flowing through two infinitely long, thin, parallel conductors one metre apart in a vacuum, which causes each to exert a force of 2×10^{-7} N on one metre length of the other.

This force may be more easily measured by using circular current-carrying coils instead of long straight wires.

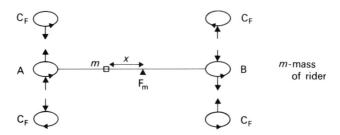

Figure 2.8 A current balance

In the apparatus in Figure 2.8, called a **current balance**, the coils C_F are fixed symmetrically above and below the movable coils A and B, and the current to the coils A and B is led in and out by the fulcrum F_m. With the current *off*, A and B are adjusted so that they are in equilibrium without the rider. When the same current is passed through all the coils, A and B are either attracted towards or repelled from the fixed coils as shown by the arrows between the coils, giving a clockwise torque, that is B going down and A going up. The rider is moved along a distance x to re-establish equilibrium.

At balance the moment of the force $mgx \propto I^2$ where the constant of variation depends only on the geometry of the apparatus, hence the current is being obtained from non-electrical measurements. Ideally, the experiment should be carried out in a vacuum.

2:B.2 *Resistance*

Unit: ohm (Ω), scalar.
The resistance* of a wire or piece of a substance is a measure of the opposition to the current flowing through it when a potential difference is applied.

In the gravitational case, a ball will accelerate under gravity if dropped in a vacuum, but will reach a steady velocity if dropped into a jar of viscous liquid due to the opposition to its movement of the liquid. Similarly, the electrons may be accelerated between two electrodes in a cathode ray tube but experience resistance when travelling down a wire.

Ohm's Law

Experiments with many common conductors show that if the potential difference between two points is increased the current increases in direct proportion, providing physical properties such as temperature remain constant.

---

* Notice that as a general rule a word ending in:
  -or   represents a **device**   e.g. resistor
  -ance represents a **property** of the device   e.g. resistance
  -ivity represents a property of the **material** e.g. resistivity

$$V \propto I \quad \text{or} \quad \frac{V}{I} \text{ is a constant at a given temperature}$$

This constant is called the resistance.

$$V = IR$$
$$V \quad A \quad \Omega$$

Note: The potential difference V is *between* two points, the current I flows *through* the material, and the resistance R is the total resistance of the material between those points.

Good conductors, such as metals, obey Ohm's law and are termed *ohmic* or *linear*. A graph of I against V will give a straight line.
Semiconductors including thermistors usually *do not* obey Ohm's law and are termed *non-linear* or *non-ohmic*.

Definition of the ohm

If a potential difference of one volt applied between two points produces a current of one ampere then the resistance of the material between those two points is one ohm.

The **conductance** is the reciprocal of the resistance; conductance $= \dfrac{1}{R}$, with a unit of ohm^{-1} or siemens S.

Factors affecting the resistance of a resistor

A **resistor** is a piece of material, usually in the form of a wire, which produces a certain opposition to the current.

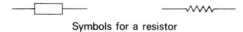

Symbols for a resistor

A long piece of wire will have more opposition than a short piece, similarly a thin wire has a greater opposition than a thick wire. Accordingly the resistance of a resistor will depend on its shape, size, and the material from which it is made.
It can be shown that:

$$R \propto l, R \propto \frac{1}{A}, \text{ and } R \propto \text{ resistivity of the material}$$

where R is the resistance, l the length, and A the uniform area of cross section.

Resistivity

Unit: ohm metre (Ω m), scalar.

The resistivity ρ is defined from the equation

$$R = \frac{\rho l}{A}$$

where R is in Ω, l in m, A in m^2.

Thus the resistivity of a material is the resistance of a slab of the material of length one metre and uniform area of cross section of one square metre. (Notice the area may be of any shape.)

The resistivity of copper, a good conductor, is $1.7 \times 10^{-8}\,\Omega\,\text{m}$ and that of graphite, a poor conductor, is around $10^{-5}\,\Omega\,\text{m}$.

Variation of resistance with temperature

The resistance of a good conducting resistor, e.g. a metal, *increases* when the temperature rises. This property is used in the platinum resistance thermometer (see Section 6:A.1). Remember that the variation of resistance with temperature is not linear over wide ranges of temperature.

Note: When a current is passed through such a resistor, the resistor is heated, its resistance increases thereby *decreasing* the current in the circuit.

Many semiconducting substances exhibit a *decrease* of resistance with temperature (see Section 3:A.1).

A **thermistor**, a **therm**ally sensitive re**sistor**, has a large variation of resistance with temperature. Thermistors are made from semiconducting substances with controlled amounts of impurities. Many thermistors encountered in teaching laboratories show a decrease of resistance with temperature but other thermistors may exhibit an increase of resistance with temperature over certain temperature ranges.

Although this decrease of resistance with temperature is non-linear the *small size* of thermistors renders them useful as *thermometers*.

Resistors in series

R is the total resistance between A and B.
V is the total potential difference between A and B.
V_1 is the potential difference between A and M, *across* R_1.
V_2 is the potential difference between M and B, *across* R_2.
I is the current through both R_1 and R_2.

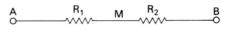

Figure 2.9 Resistors in series

Note: The current is the same at all points in any series circuit. As no current is 'lost', the currents at A, through R_1, at M, through R_2 and at B are all equal.

Remember that potential difference is the *energy* per unit charge, (see page 46). Since energy is *conserved*, then for each unit of charge (energy converted A to B) equals (energy converted A to M) plus (energy converted M to B) giving:

$$V = V_1 + V_2$$
$$IR = IR_1 + IR_2 \qquad \text{using } V = IR \text{ for each term}$$

hence
$$R = R_1 + R_2$$

The total resistance of resistors in series is the sum of the individual resistances.

Note: The potential differences *across* R_1 and R_2 may be added to give the total potential difference between A and B.

Resistors in parallel

R is the total resistance between A and B.

If V is the potential difference between A and B then

$$\text{p.d. across } R_1 = \text{p.d. across } R_2 = V$$

since the potential difference is between two points it is *not* affected by the *path* taken between those points. Similarly a mass which falls down the steep side of a hill loses as much potential energy as the same mass falling the same vertical distance down a longer moderate slope of the hill, assuming no energy loss due to friction.

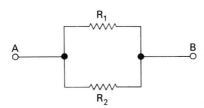

Figure 2.10 Resistors in parallel

The total current I at A will divide, part I_1 passing through R_1 and the rest I_2 through R_2. By the *conservation* of charge, the total current into a junction equals the total current leaving that junction.

At junction A
$$I = I_1 + I_2$$

$$\frac{V}{R} = \frac{V}{R_1} + \frac{V}{R_2} \qquad \text{using } V = IR$$

hence
$$\frac{1}{R} = \frac{1}{R_1} + \frac{1}{R_2}$$

> The reciprocal of the total resistance of resistors in parallel is the sum of the reciprocals of the individual resistances.

Note: For resistors in parallel, the overall resistance is always *less* than the smallest resistance. For example a $72\,\Omega$ resistor in parallel with a $24\,\Omega$ resistor produces an overall resistance of $18\,\Omega$.

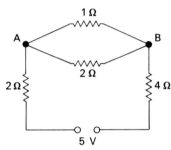

Figure 2.11

Example: Determine (a) the total resistance in the circuit, (b) the current through each resistor, (c) p.d. across the $4\,\Omega$ resistor, and (d) p.d. across the $1\,\Omega$ resistor (Figure 2.11).

(a) Total resistance $= 2 + \frac{2}{3} + 4$
$$= 6\tfrac{2}{3}\,\Omega$$

(b) Using $V = IR$ for the whole circuit,
$$5 = I \times 6\tfrac{2}{3}$$
hence $\qquad\qquad\qquad\qquad\qquad I = \tfrac{3}{4}\,\text{A.}$

This is the current through the $2\,\Omega$ and $4\,\Omega$ resistors in series with the source of 5 V. For the resistors in parallel the current will divide with twice as much taking the easier path through the $1\,\Omega$ resistor. I through $1\,\Omega = \tfrac{1}{2}\,\text{A}$ and I through $2\,\Omega = \tfrac{1}{4}\,\text{A}$.

(c) $V = IR$, p.d. across $4\,\Omega = \tfrac{3}{4} \times 4$
$$= 3\,\text{V}$$

(d) $V = IR$, either: p.d. $=$ current through $1\,\Omega$ resistor $\times$ resistance
$$= \tfrac{1}{2} \times 1$$
$$= \tfrac{1}{2}\,\text{V}$$
$\qquad\qquad$ or: p.d. $=$ current into A $\times$ total resistance A to B
$$= \tfrac{3}{4} \times \tfrac{2}{3}$$
$$= \tfrac{1}{2}\,\text{V}$$

Note: The p.d. *across* the $1\,\Omega$ resistor is the same as the p.d. *between* A and B.

Design of resistors

A resistor is expected to carry a current I and hence dissipate power of I^2R, (see page 66). The maximum power a resistor can take is called its **power rating**. Absorption of power may lead to a rise in temperature which can influence the design.

Many small cylindrical shaped resistors have a series of coloured bands to indicate their resistance value and tolerance. The **tolerance** is the percentage accuracy of the indicated value.

Example: The colour code of resistors is given below.
The first band indicates the first significant figure of the resistance value.
The second band indicates the second significant figure of the resistance value.
The third band indicates the multiplier, i.e. $\times 10^N$.
The fourth band indicates the tolerance if under 20%.
Colours correspond to digits as follows:

Colour:	black	brown	red	orange	yellow	green	blue	violet	grey	white
N:	0	1	2	3	4	5	6	7	8	9

with tolerance colours of silver 10%, gold 5%, red 2%, brown 1%.

Typical cylindrical resistor dimensions are around 1.7 mm diameter and 1.4 mm length for a 0.125 W rating, up to about 9 mm in diameter and 24 mm long for a 2 W rating.

(1) Wire wound resistor

The small *fixed* value resistors have wire wound on a hollow ceramic or ground glass tube, with dimensions in the range quoted above. Resistors of 1% tolerance can be made using nichrome wire. Manganin wire is used for even higher precision, since its resistance varies little with temperature.

The larger *variable* wire wound resistors have a sliding contact which moves across the windings. The largest types are cylindrical, about 30 cm in length, with power ratings of a few hundred watts. Other smaller types have the wire wound on a doughnut-shaped former with the sliding contact connected to a central spindle. The resistance value is determined by the length, thickness, and resistivity of the wire used.

(2) Carbon composition resistor

These carbon resistors are made by mixing finely powdered carbon with a non-conductor and enclosing the mixture in a cylindrical tube. The outer surface is coated with an insulator. These resistors have a tolerance of about 10%, and are colour coded with dimensions in the range quoted above.

(3) Thin film carbon resistors

A thin film of carbon is deposited on a ceramic tube and covered by a coating of baked enamel which protects the resistor and helps to conduct away unwanted heat. These resistors are colour coded, with a tolerance of around 2%.

(4) Metal film and metal oxide resistors

Thin films of metal, often nickel–chromium alloys, or metal oxide, are deposited on a ceramic former. The metal film may only be a few millionths of a centimetre thick. These are also colour coded with dimensions in the quoted range.

Note: For these carbon and film resistors the value of the *resistance* is altered by increasing or decreasing the *amount* of conducting material.

Uses of resistors

(1) A resistor *limits* the current flowing, the larger the resistance the smaller the current.
(2) A resistor in a circuit has a p.d. across it equal to IR. In a series circuit

containing two resistors R_1 and R_2 where $R_1 > R_2$, the p.d. across R_1 will be the larger. Thus a resistor can be placed in a circuit with the function of *providing* a desired potential difference.

(3) A resistor can *convert* electrical energy to heat or other forms of energy.

2:B.3   *Electromotive Force*

A **cell** is a chemical device which imparts energy to the electrons passing through it so maintaining a potential difference between its terminals.

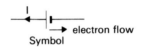

Symbol   → electron flow

The long line indicates the positive terminal hence the conventional direction of the current is in the direction shown above. The actual electron flow is in the *opposite* direction. Since electric field strength, a vector quantity, is defined in terms of direction a *positive* charge would move, the direction of current flow is by *convention* from the positive to the negative terminal in the external circuit. Remember that current is a scalar quantity and this theoretical 'direction' of the current is only of interest when studying electromagnetic effects.

Internal resistance of a cell or source

When a cell is supplying a current to an external circuit the electrons are continually passing through the cell. There is a certain opposition to this current inside the cell, which is termed the **internal resistance** of the cell.

Electromotive force, e.m.f.

Unit: volt, scalar.

> The e.m.f. is the energy a cell or source imparts to each unit of charge passing through it.

An e.m.f. of one volt implies that each coulomb of charge acquires an energy of one joule.

When a cell is supplying current to an external circuit, some energy is lost due to the internal resistance of the cell. Hence the potential difference between the terminals of the cell, the **terminal potential difference, t.p.d.**, will be less than the e.m.f.

e.m.f., ε, is equivalent to the p.d. between the terminals when the cell or source is *not* supplying an external current, i.e. an open circuit.

t.p.d., V, is the p.d. between the terminals when the cell or source *is* supplying a current I.

The difference between ε and V, due to the energy lost inside the cell because of its internal resistance, is termed the **lost volts**.

$$\text{Lost volts} = \varepsilon - V$$

Note: An electrical source or supply is equivalent to an e.m.f. together with a series resistor, which is the internal resistance r.

Cells in series and parallel

If cells are placed in **series**, the total e.m.f. is the sum of the individual e.m.f.s, providing the polarity of the connections is taken into account. Each electron acquires more energy as it passes through successive cells, see Figure 2.12.

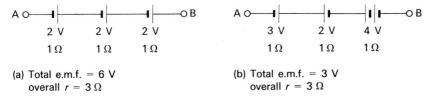

(a) Total e.m.f. = 6 V
    overall r = 3 Ω

(b) Total e.m.f. = 3 V
    overall r = 3 Ω

Figure 2.12   Cells in series

For both these arrangements the total internal resistance will be the *sum* of the individual internal resistances since all the electrons pass through *all* the cells.

For *similar* cells in **parallel** each electron will pass through one cell only. Therefore the total e.m.f. will be the e.m.f. of one cell.

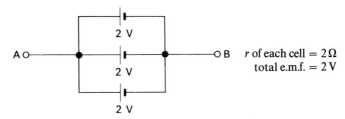

    r of each cell = 2 Ω
    total e.m.f. = 2 V

Figure 2.13   Cells in parallel

But the total internal resistance r_T will be given by $1/r_T = 1/r_1 + 1/r_2 + 1/r_3$ that is $r = \frac{2}{3}$ Ω, the internal resistance of each cell being 2 Ω. In this case all the electrons do *not* pass through all the cells.

The complete circuit

In Figure 2.14, r is the internal resistance and R the total external resistances. For a complete circuit, energy must be *conserved*.

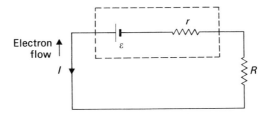

Figure 2.14   Cell with internal resistance r

For unit charge:

energy supplied by cell or source = energy converted by the external
resistances R and by the
internal resistances r

$$\text{e.m.f. of supply} = IR + Ir \qquad\qquad (\text{using } V = IR)$$

$$\text{e.m.f.} = \text{t.p.d.} + \text{lost volts}$$

Note: The lost volts depend on the current in the external circuit as well as the actual internal resistance of the cell.

Example: Apply the conservation of energy to determine a relationship between the e.m.f.s and the individual resistances for the circuit shown in Figure 2.15.

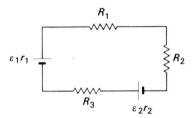

Figure 2.15 Series circuit

Energy supplied per unit charge = energy converted by resistors/unit charge
sum of e.m.f.s = sum of p.d.s across the resistors
sum of ε = sum of (IR) + sum of (Ir)
$$\varepsilon_1 + \varepsilon_2 = IR_1 + IR_2 + IR_3 + Ir_1 + Ir_2$$

Thus in general:

$$\text{sum of } \varepsilon = I \,(\text{total external } R + \text{sum of internal } r)$$

assuming that the *supplies* are in series with each other.

Use the equation
$$\varepsilon = I(R + r)$$ for a *complete* circuit where R is the *total* external resistance.

and

$$V = IR$$ between *two points* where V is the p.d. *between* those points and R the resistance between those points.

Note: The word *voltage* is a term that is applied indiscriminately to both e.m.f. and p.d., since both these quantities are measured in volts. Care should be taken *not* to *confuse* the two quantities.

Example: Using the circuit shown in Figure 2.16, determine (*a*) the current through the $5\,\Omega$ resistor, (*b*) the p.d. across the $1\,\Omega$ resistor, and (*c*) the t.p.d. of the 4 V source.

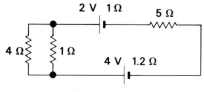

Figure 2.16

In most problems the total current is determined first.
Total resistance of $4\,\Omega$ and $1\,\Omega$ in parallel $= 0.8\,\Omega$

$$\text{using } \varepsilon = I(R+r)$$
$$4-2 = I(5.8+2.2)$$
$$I = 0.25\,\text{A}$$

(*a*) This total current flows through the $5\,\Omega$ resistor

$$I\,(\text{through } 5\,\Omega) = 0.25\,\text{A}$$

(*b*) Using $V = IR$ across the $4\,\Omega$ and $1\,\Omega$ combination
$$V = 0.25 \times 0.8 \text{ using the } total\ I \text{ and } total\ R$$
$$\text{p.d. across } 1\,\Omega = 0.2\,\text{V}$$

(*c*) t.p.d. $=$ e.m.f. $-$ lost volts
$$= 4-(1.2 \times 0.25)$$
$$= 3.7\,\text{V}$$

Experiment 2.2 *Measurement of internal resistance*

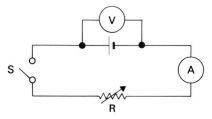

Figure 2.17

The potential difference across the terminals of the cell in an *open circuit* is equal to the e.m.f. Thus with S open the voltmeter will record the e.m.f. with reasonable accuracy. In practice, the value recorded will be slightly too low, because of the finite current flowing through the voltmeter, see page 75.

S is then closed. The current I is recorded on the ammeter and the p.d. V across the resistor is recorded on the voltmeter for different values of resistance R. After each reading the cell is left off for a few minutes before the next reading is taken

Using $\varepsilon = I(R+r)$, then $\varepsilon = V+Ir$ where V is the p.d. across R, giving $\varepsilon - V = Ir$. Hence r can be calculated from each pair of readings V and I, using the above measured value of e.m.f.

Potential divider

It is sometimes useful to be able to obtain a *range* of potential differences from one supply. Consider a 10 V supply of negligible internal resistance connected to a series of 3 kΩ resistors as shown in Figure 2.18.

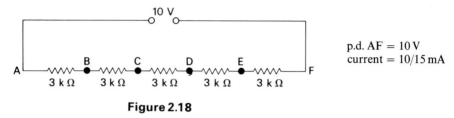

p.d. AF = 10 V
current = 10/15 mA

Figure 2.18

The p.d. AB = 2 V; p.d. AC = 4 V; p.d. AD = 6 V; p.d. AE = 8 V.

This series of resistors can *divide* the 10 V applied p.d. into 2 V intervals. By an extension, ten equal resistors could divide this 10 V into 1 V intervals, or a resistor with a movable contact could provide *any* desired interval.

Example: A 24 V supply is connected to three resistors as shown in Figure 2.19.
   R is a 6 Ω resistor with a movable contact. State the positions in the circuit which give p.d.s of 4 V, 12 V, and 15 V with respect to the point O.

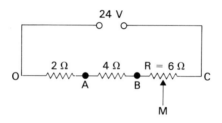

Figure 2.19

The total resistance is 2+4+6 = 12 Ω. Thus there is a p.d. of 2 V per ohm from the point O. Hence OA has a p.d. of 4 V. OB has a resistance of 6 Ω, that is a p.d. of 12 V. For 15 V a resistance of 7.5 Ω is required, thus the movable contact on resistor *R* should be 1.5 Ω from B to give the p.d. OM = 15 V.

Note: A potential divider can be constructed by placing a resistor with a *movable* contact across the supply *V* (see Problem 2.11). A p.d. which is *variable* from zero to a maximum value *V* can be obtained. A potential divider providing *fixed* p.d.s has two or more fixed value resistors connected across the supply.

   It is important to remember that these values are the p.d.s which *could* be supplied across another component or resistor. If that resistor is joined into the circuit, and current flows then the original circuit will have been *changed*.

   A **potentiometer** is a device for measuring an e.m.f. or p.d. by *balancing* the unknown p.d. against a known p.d. For example, consider an external p.d. of 4 V connected across OA of Figure 2.19, via a galvanometer, before the connection to A. It would be observed that *no* current would flow through the galvanometer, since the two p.d.s are balanced. A commercial potentiometer requires an initial standardisation procedure. Then a set of dials are adjusted until no current flows through a sensitive galvanometer. The reading on the dials indicates the value of the p.d. or e.m.f. under test.

2:B.4 *Electrical Energy*

From the definition of potential difference, see page 46,

$$\text{Energy} = \text{potential difference} \times \text{charge}$$
$$= VQ$$

$$\text{Energy} = VIt$$
$$\underset{\text{J}}{\Big|} \quad \underset{\text{V}}{\diagup} \quad \underset{\text{A}}{\diagdown} \quad \underset{\text{s}}{\diagdown}$$

Also using $V = IR$ and substituting one obtains

$$\text{Energy} = I^2 Rt$$
$$= \frac{V^2}{R} t$$
$$\diagdown \Omega$$

Electrical energy may be converted into other forms of energy, for example, mechanical energy in a motor, light energy in a light bulb, heat in an electric fire.

Example: An electrical heating element of resistance $7\,\Omega$ is immersed in 0.1 kg of water at 20°C for 3 minutes. If the current is 4 A, what is the final temperature of the water? (The specific heat capacity of water is $4.2 \times 10^3\,\text{J kg}^{-1}\,\text{K}^{-1}$.) State any assumptions that are made.

$$\text{Electrical energy} = I^2 Rt = 16 \times 7 \times 60 \times 3\,\text{J}$$
$$\text{Heat taken in by the water} = \text{mass} \times \text{specific heat} \times$$
$$\text{temperature rise}$$
$$= 0.1 \times 4.2 \times 10^3 \times \text{temp. rise}$$

Assuming: (*a*) all the electrical energy is converted into heat,
　　　　　(*b*) all this heat is absorbed by the water with no heat loss.

Then　　　　　Electrical energy = heat taken in by the water
$$16 \times 7 \times 60 \times 3 = 0.1 \times 4.2 \times 10^3 \times \text{temp. rise}$$

Which gives a temperature rise of 48°C.
Final temperature of the water = 68°C.

A **joulemeter** is an instrument which may measure electrical energy directly.

2:B.5 *Power*

This is the rate of expending energy, or the rate of energy transfer, which in the electrical case will be given as

$$\text{Power} = \frac{\text{energy}}{\text{time}} = \frac{IVt}{t}$$

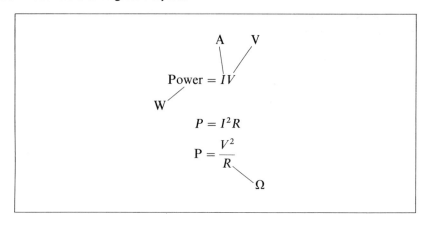

A *source* of electrical energy can be represented by an e.m.f. ε and internal resistance *r*. The part of the circuit which utilises the electrical energy is called the **load**, e.g. a light bulb, a motor, or a resistor. Thus the load converts electrical energy into other forms, e.g. heat or mechanical work done. The *rate* at which the load *converts* electrical energy is equal to $V_L I_L$ where V_L is the p.d. across the load.

The amount of *power transferred* to a load will depend on the values of the internal resistance of the source *r* and the resistance of the external circuit *R*.

Example: A 40 V source, of internal resistance 8 Ω, is connected to external devices of resistance *R*. State for each value of *R*, the current, the p.d. across the load of resistance *R*, the power supplied by the source, and the power transferred to the external resistance *R* if
(*a*) $R = 2\,\Omega$, (*b*) $R = 8\,\Omega$, (*c*) $R = 32\,\Omega$,
(*d*) Comment on the power transferred to the external circuit.

The calculation for part (*a*) is shown in full, but those for (*b*) and (*c*) are left as an exercise for the reader to verify.
Using $\varepsilon = I(R+r)$, $40 = I(2+8)$ giving $I = 4\,A$.
Using $V = IR$ the p.d. across R, $V_R = 4 \times 2 = 8\,V$.
Since $P = IV_S$ where V_S is 40 V, the power supplied $P_S = 160\,W$
Using $P = IV_R$ where V_R is the p.d. across R
the power transferred to the resistor R is $4 \times 8 = 32\,W$.
The results are summarised below:

	(*a*)	(*b*)	(*c*)
Current I (A)	4	2.5	1
p.d. across R, V_R (V)	8	20	32
Power supplied, P_S (W)	160	100	40
Power transferred P_t (W)	32	50	32

(*d*) The power transferred to the external circuit is the largest when $R = 8\,\Omega$.
(But it must be noticed that when $R = 32\,\Omega$, i.e. $R \gg r$, the p.d. across R is at a maximum.)

Note: The maximum *transfer* of power occurs when the internal resistance of the source is *equal* to the external resistance in the circuit.

Experiment 2.3 *Transfer of power between source and load*

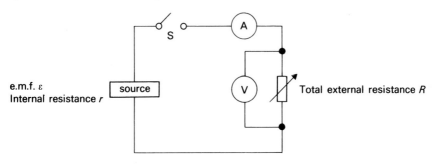

e.m.f. ε
Internal resistance r

Total external resistance R

Figure 2.20

With S closed, the current I through R, and the p.d. V across R, are measured for different values of R and the power transferred P_t calculated for each pair of results. S is opened and the voltmeter records the e.m.f. of the source ε. The internal resistance of the source is determined using $\varepsilon - V = Ir$. The results show maximum power transfer when $R = r$. If the source is a battery, readings of V and I should be taken quickly to avoid the battery running down.

When purchasing electricity one must pay for the power and the time period for which it is used. The unit employed is the kilowatt-hour, that is a power of one kilowatt used for one hour will have a certain cost.

Note: The kilowatt-hour has the same dimensions as energy, being a power multiplied by a time.

Example: One kilowatt-hour on a certain meter costs 3p. How much does it cost to use a 10 A fire off 240 V mains for $1\frac{1}{2}$ hours?

$$\text{Power used} = 10 \times 240 \text{ W}$$
$$= 2.4 \text{ kW}$$
$$\text{Number of kilowatt-hours} = 2.4 \times 1\frac{1}{2}$$
$$= 3.6$$

Thus
$$\text{Cost} = 3.6 \times 3$$
$$= 10.8 \text{ p}$$

Problems

2.6 A source has an internal resistance r of 1Ω. The current through the 3Ω resistor in Figure 2.21 is 2 A, determine: (*a*) the p.d. across the 1.5Ω resistor, (*b*) the current through the 2Ω resistor, (*c*) the current through the source, and (*d*) the lost volts and e.m.f. of the source.

2.7 The heating element in an electric kettle is rated at 2 kW. How long does it take to bring 500 g of water at 20°C to the boil if the specific heat capacity of water is $4.2 \times 10^3 \text{ J kg}^{-1} \text{ K}^{-1}$? Assume 10% of the heat is lost to the kettle and the surroundings.

2.8 How much will it cost to use a 60 W light bulb off 240 V mains for 2 hours if the cost of electricity is 3p per kilowatt-hour? What is the current flowing through this bulb?

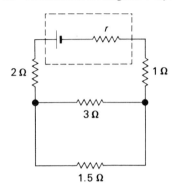

Figure 2.21

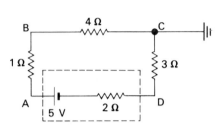

Figure 2.22

2.9   Three cells, each of e.m.f. 2.6 V and internal resistance 1 Ω, are supplying a current to an external resistor of 4 Ω. Calculate the current through the resistor when: (a) the cells are in series, and (b) the cells are in parallel.

2.10 A cell of e.m.f. 5 V and internal resistance 2 Ω is connected to three resistors as shown in Figure 2.22. Point C is earthed (i.e. its potential is zero). Calculate the p.d. between B and C, the terminal p.d. across the cell, and state which point has a potential of $-1\frac{1}{2}$ V.

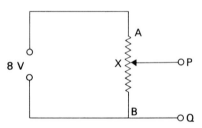

Figure 2.23

2.11 A 20 cm long variable resistor AB, of uniform resistance per unit length and a total value of 100 Ω, is connected to an 8 V supply of negligible internal resistance. Using Figure 2.23 determine the length of resistance AX to provide a p.d. across PQ of:
(a) 4 V   (b) 2 V   (c) 1 V.
(d) State the potential gradient of AB in V cm^{-1}.
(e) With X at the 10 cm mark, what would happen to the p.d. across AX if a resistor of 200 Ω was connected to PQ?

2.12 A source of 12 V of negligible internal resistance, and standard resistors of 30 Ω, 40 Ω, and 50 Ω are available. Draw a circuit diagram using these three resistors to construct a potential divider. State the possible values of fixed p.d.s which are available.

2.13 A source, of e.m.f. of 12 V and internal resistance 2 Ω, is connected in a series circuit with a 3 Ω and 5 Ω resistor. Determine: (a) the p.d. across each resistor (b) the rate of energy transfer in each resistor, (c) the percentage of the total power transferred to the 5 Ω resistor.

2.14 A cell of e.m.f. ε and internal resistance r is connected to an external resistor R as shown in Figure 2.24. The resistor R is altered and a series of readings of V and I for each setting of R are shown below.

p.d. V (volts)	0.8	1.0	1.1	1.2	1.3
current I (A)	0.81	0.56	0.44	0.31	0.19

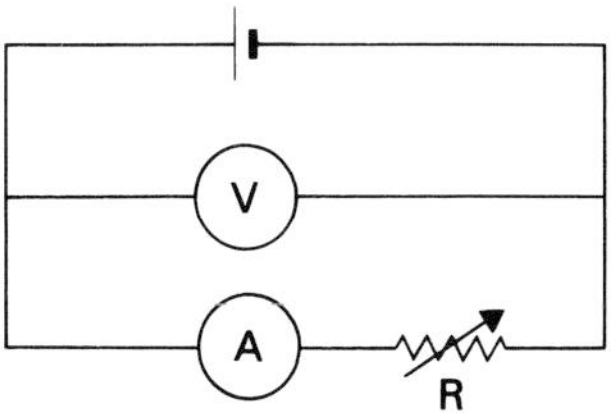

Figure 2.24

(*a*) Show that $\varepsilon - V = Ir$ for this circuit. (*b*) Plot a graph of V and I using these results. (*c*) Determine ε and r from the graph. (*d*) How could the e.m.f. have been measured directly using the circuit shown? Comment on the accuracy of the result so obtained.

C: Magnetic Effects of a Current, the Galvanometer, and Resistance Measurement

2:C.1 *Magnetic Field associated with a Current*

A *stationary* charge produces an *electric* field in the region around it.
A *moving* charge also produces a *magnetic* field in that vicinity.
Thus any current will have an associated magnetic field.

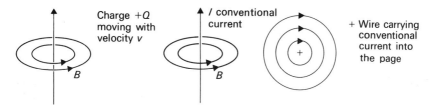

Figure 2.25 Lines of magnetic field

The *lines* of magnetic field, called lines of **magnetic flux**, show the *direction* and *distribution* of the magnetic field.

For a linear stream of charges, or a single, straight, current-carrying conductor, these lines are circular around the conductor but in a plane perpendicular to the direction of the current I. The *direction* of the magnetic field B around the conductor may be determined by the **right-hand corkscrew rule**.* Using the right hand, the thumb points in the direction of the current and the fingers, clenched round the thumb, will give the direction of the field.

With a solenoid a uniform magnetic field may be obtained along its axis, as shown in Figure 2.26 on the following page.

---

* For electron flow use the *left* hand, with the thumb in the direction of the electron flow.

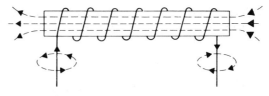

– – – – Lines of magnetic field

Figure 2.26 The magnetic field of a solenoid

The external field of a solenoid is similar to the external field of a bar magnet. The direction of the field may be determined by applying the corkscrew rule to the end turn of the solenoid wire.

The *magnitude* or strength of the magnetic field B at any point will depend on:

the value of the current I, or rate of flow of charge Q,

the distance of the point from the conductor, and

the medium.

For example, the magnetic field of a solenoid is increased if a cylinder of soft iron is placed inside. Notice that the magnitude of the magnetic field due to a straight conductor *decreases* on moving away from the conductor, but does *not* alter on moving up and down parallel to that conductor.

A permanent magnet is made from certain ferromagnetic materials such as steel, or alloys of iron, nickel, and cobalt, such as alnico. The circulation and spinning *movement* of the electrons in the atoms are the cause of this magnetism.

Interaction of two magnetic fields

Two magnetic fields can cause a **force** of attraction or repulsion. Combinations of permanent magnets and current carrying conductors will now be considered. In Figures 2.28 to 2.32, the conventional current direction is indicated.

Note: Adjacent lines of magnetic field in *opposing* directions cause a force of *attraction.*

(1) **Two permanent magnets** exert a force on each other.

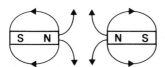

Lines of magnetic field in the same direction between the magnets

Figure 2.27 Repulsion of two bar magnets

Two north poles result in repulsion of the magnets, but a north pole and a south pole attract each other.

(2) **Two current-carrying conductors** produce a pair of forces.

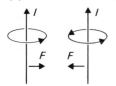

Lines of magnetic field in opposing directions between the conductors

Figure 2.28 Attraction of two conductors

The ampere is defined in terms of this effect, see page 53.

(3) One magnet and one current-carrying conductor often give rise to movement of the conductor. The lines of magnetic field tend to straighten out, giving the force on the conductor shown in Figure 2.29

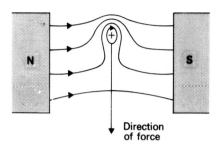

Direction
of force

Lines of magnetic field above the conductor are in the same direction

Figure 2.29 Force on a conductor in a magnetic field

Fleming's left-hand rule* may be used to determine the direction of the force. The forefinger, second finger, and thumb are held mutually perpendicular to each other with:

> Forefinger in the direction of the field
> Second finger in the direction of conventional current
> Thumb gives the direction of the motion

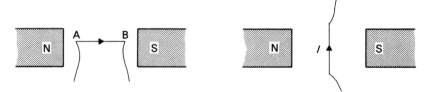

(a) Force on the conductor AB is zero

(b) Force on the conductor is into the page

Figure 2.30 Examples of Fleming's left-hand rule

For the *particular* case of a straight conductor placed perpendicular to a uniform field, the force F depends on:

> the magnitude of the magnetic field B,
> the length of the conductor in the field, and
> the value of the current flowing, I.

These last two, length and current, are equivalent to the amount of charge flowing Q, with velocity v.

Magnetic induction or magnetic flux density †

Unit: tesla (T), vector.
The strength of the magnetic field is termed **magnetic induction** or **magnetic flux density** and is denoted by the letter B.

The strength of the electrostatic field is defined by the force on a unit

---

* For electron flow use the *right* hand, with the second finger in the direction of the electron flow.

† Details of magnetic induction and the tesla are not required for the Higher Grade Physics examination.

stationary positive charge with units of NC^{-1}. By analogy the strength of the magnetic field can be defined as the force on a unit 'charge moving perpendicular to the field', *or* as the force on a unit 'current-carrying conductor' placed perpendicular to the field, i.e. the force on a 1 m conductor carrying a current of 1 A. This gives the unit of *B* as $NA^{-1}m^{-1}$ which is called the **tesla**.

Note: The force depends on the *orientation* of the conductor, or moving charges, in the field.

   If the magnetic flux density *B* is large then this can be *represented* by lines of magnetic flux drawn close together.

Torque on a current-carrying wire loop in a magnetic field

Figure 2.31 shows a single loop of wire ABCD between the poles of a magnet. From the directions of the field and the conventional current, the force on AB is outwards from the plane of the paper while that on CD is downwards into the

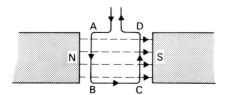

Figure 2.31   Current-carrying loop of wire in a magnetic field

paper. This results in a turning effect, a **torque**, tending to turn the coil through a right angle. After turning through a right angle the forces would still be the same in magnitude but would be acting in opposite directions along the same straight line so that their torque would have become zero. Thus the coil will turn until the plane of the coil is perpendicular to the page.

   The magnitude of the forces producing the torque will depend on the value of the current flowing in the loop, and the number of turns or loops. This effect is used in the *moving coil galvanometer*.

The d.c. motor

This uses a coil suspended between the poles of a magnet, but now continuous rotation is required. Current is fed into the coil *via* two half circles of metal,

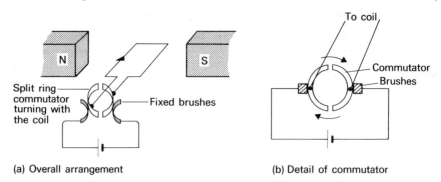

(a) Overall arrangement   (b) Detail of commutator

Figure 2.32   A d.c. motor

called a **split-ring commutator**, mounted on the shaft carrying the coil. This arrangement automatically reverses the current after the loop has turned through a right angle and each successive half revolution thereafter. The coil is therefore kept constantly rotating.

2:C.2 *The Principle and Use of the Moving-coil Galvanometer*

One of the most common meters used to measure p.d. and current is the **moving-coil galvanometer**. This instrument has a coil, mounted on a spindle, and suspended between the pole pieces of a permanent magnet. The current enters and leaves the coil at X and Y through the springs. Jewelled bearings carry the coil pivots. When a current is passed through the coil, the magnetic field produced by this current interacts with the magnetic field of the permanent magnet to give a force on either side of the coil.

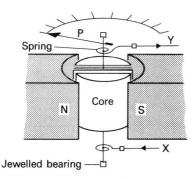

Figure 2.33 Moving-coil galvanometer

Thus a *torque* is produced and the coil tends to *rotate*. This turning effect is resisted by springs attached to the spindle and a pointer P is deflected across a scale. Using *curved* pole pieces the deflection of the pointer can be made directly proportional to the current passing through the coil, giving a *linear* scale on the meter.

Note: Currents of the order of a few mA may be measured by this instrument. Larger currents will tend to damage the galvanometer.

Use of a galvanometer to measure currents of several amperes

To use a galvanometer as an ammeter, a resistor, called a **shunt**, is placed in *parallel* with the galvanometer.

Example: A certain galvanometer has a resistance of $10\,\Omega$ and a maximum permitted current of 6×10^{-3} A, that is a current of 6 mA will give a full-scale deflection, (f.s.d.). How may currents of up to 2 A be measured?

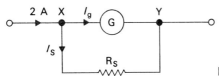

Figure 2.34 Galvanometer with shunt R_S

A resistor R_S, termed a shunt, must be placed in *parallel* with the galvanometer so that only a small fixed fraction of the total current passes through the galvanometer.

For a maximum current of 2 A and f.s.d. of 6×10^{-3} A,

$$I_g = 6 \times 10^{-3} \text{ A}$$
$$I_S = 2 - 6 \times 10^{-3}$$
$$= 1.994 \text{ A}$$

The p.d. XY across the galvanometer is equal to the p.d. XY across the resistor R_S. Using Ohm's Law between X and Y,

$$\text{p.d. XY} = I_g R_g = I_S R_S$$
$$6 \times 10^{-3} \times 10 = 1.994 \times R_S$$
$$R_S = 0.03 \, \Omega$$

The value of the shunt depends on the currents to be measured, the larger the current the smaller the value of the shunt resistance, hence more current may then bypass the galvanometer.

Use of a galvanometer as a voltmeter to measure a large p.d.

For use as a voltmeter a resistor, called a **multiplier**, is placed in *series* with the galvanometer.

Example: Using the same galvanometer as in the previous section, with f.s.d. 6×10^{-3} A and resistance $10 \, \Omega$, implies a maximum p.d. across the voltmeter of $6 \times 10^{-3} \times 10$ equal to 6×10^{-2} V.

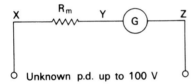

Unknown p.d. up to 100 V **Figure 2.35** Galvanometer with multiplier R_m

Now a resistor R_m, termed a multiplier, must be placed in *series* with the galvanometer in order that most of the external p.d. is dropped across R_m.

Consider a numerical example. How can this same meter be used to read up to 100 V?

For a maximum p.d. XZ of 100 V and p.d. YZ of 6×10^{-2} V, the current through the galvanometer is 6×10^{-3} A, which is also the current through R_m.

Using Ohm's Law between X and Z, $V = IR$

$$100 = 6 \times 10^{-3} (R_m + 10)$$
$$R_m = 1.67 \times 10^4 \, \Omega$$

Observe that the resistance of the multiplier is much larger than that of the galvanometer.

Note: In general, a shunt has a low resistance and a multiplier has a high resistance.

2:C.3 The Effect in a Circuit of the Resistance of the Galvanometer

The galvanometer as an ammeter in a circuit

In order to *measure* the current in a circuit, an ammeter (galvanometer plus shunt) must be introduced into the circuit. The circuit has then been *altered*.

Example: Determine the current flowing in the circuits of Figure 2.36.

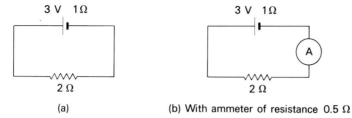

(a)   (b) With ammeter of resistance 0.5 Ω

Figure 2.36

The only difference in the two circuits is the inclusion of the ammeter in Figure 2.36(b).

For (a) $\varepsilon = I(R+r)$ For (b) $\varepsilon = I(R+r)$
          $3 = I \times 3$ $3 = I(2+0.5+1)$
          $I = 1\,\text{A}$ $I = 0.86\,\text{A}$

The *calculated* value of the current is 1 A. When the ammeter is placed in the circuit it will *record* a value of 0.86 A, because the circuit has been altered. The percentage error is 14%.

The above example shows that it is desirable for the resistance of the ammeter to be as *low* as possible. If the resistance of the ammeter had been 0.1 Ω the current recorded would be 0.97 A and the error only 3%.

The galvanometer as a voltmeter in a circuit

To *measure* a p.d. a voltmeter (galvanometer plus multiplier) must be placed in the circuit. Again the inclusion of the voltmeter will *alter* the circuit.

Example: In order to measure the e.m.f. of a 4 V cell, a voltmeter of resistance 250 Ω is connected across the terminals. Calculate the percentage error if the internal resistance of the cell is 2 Ω.

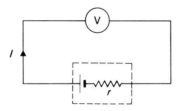

Figure 2.37   Voltmeter to measure e.m.f.

Using $\varepsilon = I(R+r)$ gives $4 = I \times 252$ and $I = 4/252\,\text{A}$
This is the current flowing through the cell and the voltmeter.

$$\text{p.d. recorded on the voltmeter} = \frac{4 \times 250}{252} = 3.97\,\text{V}$$

$$\text{percentage error} = 0.75\%$$

This error can be reduced by choosing a voltmeter of *large* resistance, but cannot be eliminated.

Example: Compare the calculated p.d. across the 300 Ω resistor with the value observed

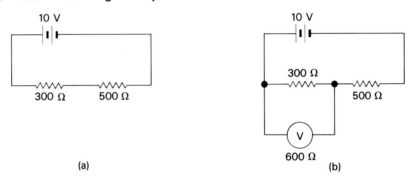

Figure 2.38 Voltmeter to measure p.d.

on a voltmeter placed across that resistor, as shown in Figure 2.38. The internal resistance of the supply is negligible.

Without the voltmeter
using $\varepsilon = I(R + r)$ then $10 = I \times 800$ and $I = 0.0125$ A
giving the p.d. across the $300\,\Omega$ resistor $= 300 \times 0.0125$
$$= 3.75\text{ V}$$
When the voltmeter is included the total resistance of the circuit changes, Figure 2.38(b).
Total resistance $=$ (resistance of $300\,\Omega$ and $600\,\Omega$ in parallel) $+ 500\,\Omega$
$$= 700\,\Omega$$
Using $\varepsilon = I(R + r)$ then $10 = I \times 700$ giving $I = 0.0143$ A
Using $V = IR$ for AB with the total current into AB and the total resistance of AB

$$\text{p.d. across } 300\,\Omega \text{ resistor} = 20 \times 0.0143$$
$$= 2.86\text{ V}$$

This example shows that the measured p.d. is too *low*. The situation would be improved if the resistance of the voltmeter was *larger*.

Note: When an ammeter or voltmeter is included in a circuit that circuit is altered. It is desirable to have an ammeter of low resistance and a voltmeter of high resistance.

2:C.4 *Methods of Measuring a Resistance R$_x$*

(1) Ammeter—voltmeter method

In Figure 2.39(a), the ammeter will record the current and the voltmeter the potential difference across R_x. By using Ohm's Law, $V = IR$, the value of the resistance may be calculated. However, although the voltmeter measures the p.d. across R_x correctly, the ammeter does *not* record the current through R_x but the total current entering P which is the sum of the currents through R_x and the voltmeter. If the resistance of the voltmeter were the same as that of R_x, then only half the measured current would pass through R_x. On using $R = \dfrac{V}{I}$, I would be twice the correct value giving a 50 % error in the resistance of R_x! To obtain more accurate results a high resistance voltmeter must be employed. This method is therefore only of use in determining values of resistance which are small compared with the voltmeter resistance.

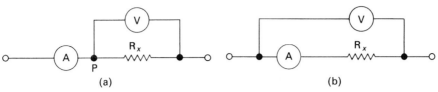

Figure 2.39 (a) Voltmeter across R_x only (b) Voltmeter across both R_x and ammeter

(2) Ammeter—voltmeter method: Alternative arrangement

To ensure an accurate measurement of the current the ammeter is placed directly in series with the resistor (Figure 2.39(b)). Unfortunately the p.d. reading on the voltmeter is now in error, being the sum of the p.d. across the ammeter as well as across R_x. For this reading to be nearly the p.d. across R_x, the p.d. across the ammeter must be small. This may be achieved by choosing an ammeter with a low resistance, compared with R_x. This is the better method for measuring large values of resistance.

(3) Substitution method—ohmmeter

This method employs a standard resistance box. With M joined to S_1 the current through the resistor R_x is noted. M is then joined to S_2, bringing the resistance box into the circuit. The resistance box is adjusted until the *same* current as before is obtained. The value of the resistance on the box is then equal to the

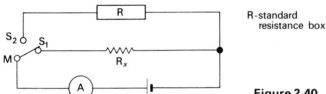

R-standard
resistance box

Figure 2.40 Substitution method

value of the unknown resistance R_x. As the e.m.f. of the source may vary slightly during the experiment, the current recorded with R_x in the circuit should be checked again on completing the observations. This method may be used over a large range of resistances but is limited by the steps with which the resistances in the box may be altered.

With the standard resistance box in the circuit the ammeter could be calibrated in terms of resistance and then used to measure unknown resistances directly. An ammeter calibrated in such a way *together* with its *steady* supply battery is termed an **ohmmeter**. Notice that the large values of resistance will be near the zero of the ammeter's 'current scale', and observe that the resistance scale is *non-linear*.

(4) The Wheatstone bridge

The circuit is arranged as in Figure 2.41 with four resistors, R_1, R_2, R_3, and R_x. Current from the battery will flow from A to C through the two branches of the circuit.

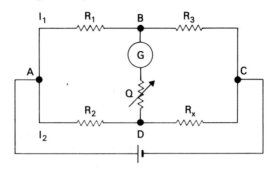

Figure 2.41 Wheatstone bridge

The resistance of one of the resistors is altered until *no current* flows through the centre zero galvanometer. The bridge is then said to be *balanced*. Because no current now flows between B and D the p.d. BD must be zero. Hence the p.d.s across R_1 and R_2 are the same and those across R_3 and R_x are also equal, giving:

$$V_{AB} = V_{AD} \quad \text{and} \quad V_{BC} = V_{DC}$$
$$I_1 R_1 = I_2 R_2 \quad \text{and} \quad I_1 R_3 = I_2 R_x$$

Dividing these equations,

$$\frac{I_1 R_1}{I_1 R_3} = \frac{I_2 R_2}{I_2 R_x}$$

$$\frac{R_1}{R_3} = \frac{R_2}{R_x}$$

or

$$R_x = \frac{R_2 R_3}{R_1}$$

Usually R_3 is a resistance box and R_1 and R_2 are resistors of known resistances in some simple ratio, for example 10:1 or 1:1000. The value of the resistance on the resistance box is varied until there is a *null deflection* on the galvanometer. The value of R_x may then be determined from the above relationship.

The method may be made accurate by using a *sensitive* galvanometer. Since only the null point is required the accuracy of the calibration of the galvanometer is unimportant. A variable resistor Q is included to protect the galvanometer. When close to the balance point the value of Q can be reduced to zero.

The ratio of $\dfrac{R_2}{R_1}$ may be varied so that the full range of the resistance box may be utilized, enabling results to four significant figures to be obtained on a resistance box with resistances up to $9000\,\Omega$. For example, if R_x is $2.574\,\Omega$ then the ratio $\dfrac{R_2}{R_1}$ may be made $1:1000$ giving a reading of R_3 as $2574\,\Omega$.

A wide range of resistances may be measured by this method, but if R_x is small the resistance of the leads in the circuit will introduce errors.

Experiment 2.4 *The Wheatstone bridge in an out-of-balance condition*

Using the circuit of Figure 2.41, R_3 is adjusted until the null point is determined and the resistor Q is set at zero. The value of R_3 is altered and readings of R_3 and current on the galvanometer recorded.

It is observed that the current flowing varies directly as the *change* in resistance value. (An example is provided by Problem 2.20 page 81.)

Note: The galvanometer current varies as the *change* of resistance, only for quite *small* changes in *one* of the resistors, from the balance point condition. (For large changes in resistance this relationship becomes non-linear.)

(5) The metre bridge

This is a special form of the Wheatstone bridge in which the resistors R_1 and R_3 are replaced by a uniform resistance wire AB, one metre in length, see Figure 2.42. Copper strip connectors are used to minimize connector resistance. A suitable resistance is removed from the resistance box R and the sliding contact touched on to the wire until the null point is obtained. As the wire is uniform, the resistances of AC and CB will be proportional to their lengths. The resistor P is included to *protect* the galvanometer when attempting to find the balance point. When near to the balance point, S is closed to short-circuit P.

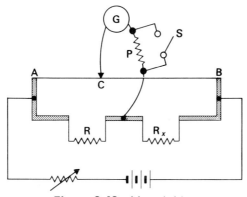

Figure 2.42 Metre bridge

Hence at balance:

$$\frac{R}{R_x} = \frac{\text{resistance of AC}}{\text{resistance of CB}} = \frac{\text{length of AC}}{\text{length of CB}}$$

Again a *sensitive* galvanometer is required for greater accuracy. The null point C should be near the centre of the wire AB so that the percentage uncertainties in measuring the lengths AC and CB are as small as possible. With the copper connecting strips errors due to the resistances of the leads are reduced, but at very small values of R_x they will cause inaccuracies.

Comparison of the five methods

Method	Range of R_x	Requirements	Errors
(1)	only small	high-resistance voltmeter	current through the voltmeter is neglected
(2)	only large	low-resistance ammeter	p.d. across the ammeter is neglected
(3)	not small	steady source	limited by the steps and range of resistance box
(4)	wide range	sensitive galvanometer	resistance of the leads when R_x is small
(5)	wide range	sensitive galvanometer, uniform wire AB, null point C near the centre of AB	resistance of the leads when R_x is very small

Problems

2.15 In the voltmeter–ammeter methods (1) and (2) for measuring a resistance R_x, would the measured values of R_x tend to be greater or less than the true values?

2.16 Using a substitution arrangement for determining the value of a resistance R_x a pupil has only one switch (see Figure 2.43). The current was noted with the switch open. The switch was then closed and the resistances on the resistance box R_B

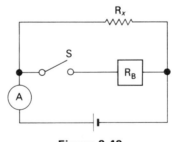

Figure 2.43

altered until the current was *twice* the previous reading. The resistance of R_x was concluded to be the same as that on the resistance box. Is this correct? Does this method have any advantages or disadvantages over method (3) described on page 77?

2.17 The e.m.f. of a 1.5 V cell of internal resistance $1\,\Omega$ is measured with a voltmeter of resistance $500\,\Omega$. The voltmeter has 0.1 V graduations. State, with reasons, what reading will be observed on the voltmeter and what conclusions concerning the e.m.f. of the cell may be drawn from this experiment.

2.18 The circuit shown in Figure 2.44 has a 20 V source of internal resistance 2 Ω.
(*a*) Calculate the current through, and the p.d. across, the 200 Ω resistor.
(*b*) Calculate the current recorded on an ammeter of resistance 20 Ω placed in the circuit.
(*c*) Calculate the p.d. recorded on a voltmeter of resistance 600 Ω placed across the 200 Ω resistor. (The ammeter does not remain in the circuit.)
(*d*) Account for the difference in the answers to (*a*) compared to (*b*) or (*c*).·

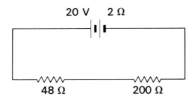

Figure 2.44

2.19 A metre bridge has a standard 1 kΩ resistor in the left hand arm and a 1.5 kΩ, nominal value, resistor in the right hand arm.
(*a*) If the balance point is 41 cm along the wire from the left hand end, is the 1.5 kΩ resistor in its marked tolerance of 10%?
(*b*) How far, and in which direction, would the balance point move if another resistor of 1.52 kΩ was placed in parallel with the nominal 1.5 kΩ resistor?

2.20

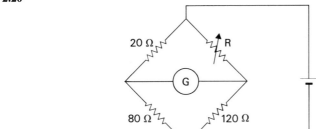

Figure 2.45

The Wheatstone bridge, shown in Figure 2.45, has a variable resistor in one arm of the bridge.
(*a*) State the value of R for a balance point.
(*b*) The reading on the galvanometer is noted when the 120 Ω resistor is altered and the following results obtained. R remains at the balance point value.

Resistance in ohm	121	122	123	124	125
Galvanometer current in mA	0.07	0.13	0.20	0.27	0.33

Plot a graph of galvanometer current against the *increase* of resistance from 120 Ω, and comment on the graph obtained.

3 Semiconductors and Alternating Current

A: Semiconductors and Transistors

3:A.1 Semiconductors

Semiconductors have a small number of mobile charge carriers and resistivities at room temperature of about 10^{-1}–$10^{3}\,\Omega\,\text{m}$.

Germanium and silicon, both of which are semiconductors, have four valence electrons per atom. In the crystal these valence electrons are linked to four adjacent atoms to give a tetrahedral lattice.

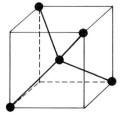

(a) Tetrahedral lattice

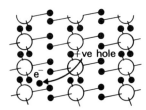

(b) Two-dimensional diagram showing the formation of a positive hole

Figure 3.1

In a pure crystal, if an electron breaks free from its position in the lattice a vacancy is left behind which has a positive charge. This position in the lattice, where there is a lack of an electron, is termed a positive **hole**. A hole may be filled by a neighbouring electron and although it is the electron which has moved in one direction the net result is a movement of the positive hole in the opposite direction. This hole may then be thought of as a **positive charge carrier**. Thus an electric current in this material consists of a drift of electrons in one direction and of positive holes in the other. This type of semiconductor is termed an **intrinsic** semiconductor. The number of electrons and positive holes are equal, being released from the lattice of the pure crystal.

If the temperature is increased more electrons are released from their lattice positions producing electron/hole pairs and hence the number of mobile charge carriers is increased. This explains the *decrease* of resistance with temperature of such materials (see Section 2:B.2).

The more important semiconducting devices are made from **extrinsic** semiconductors which consist of a pure crystal of germanium or silicon to which a very small amount of a given impurity has been added. The impurity then provides the mobile electrons or holes.

n-type silicon

This is silicon to which a very small amount of an element with *five* valence electrons (such as phosphorus, arsenic, or antimony), called a **donor**, has been added.

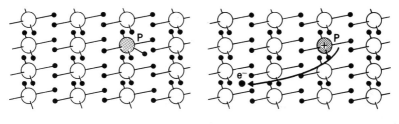

(a) Silicon with phosphorus donor   (b) The extra electron is a negative charge carrier

Figure 3.2   n-type silicon

The atoms of these impurities fit into the silicon lattice without upsetting its electrical neutrality, but because the fifth electron is not essential in the lattice structure it is free to act as a charge carrier. When this electron moves away from the parent atom it leaves, for example, a positively charged phosphorus ion, but the crystal as a whole is electrically neutral.

Note: In this n-type semiconductor the charge carriers supplied by the donor are negative electrons.

p-type silicon

This is silicon to which a very small amount of an element with *three* valence electrons (such as boron, aluminium, gallium, or indium), called an **acceptor**, has been added.

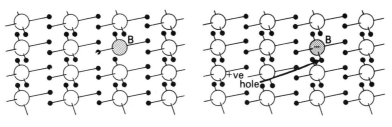

(a) Silicon with boron acceptor   (b) A positive hole on a silicon atom becomes a positive charge carrier

Figure 3.3   p-type silicon

Again the atoms of the impurity fit into the silicon lattice without affecting its electrical neutrality, but the absence of a fourth electron leaves a vacancy in the lattice. An electron from another position requires little energy to move and occupy this vacancy giving, for example, a negative boron ion and leaving behind it a positive hole. This hole is then free to be moved about the crystal by other lattice electrons, becoming a positive charge carrier.

Note: In the p-type semiconductor the charge carriers supplied by the acceptor are positive holes.

The donor or acceptor concentrations are about one part in 10^8.

Similar n-type and p-type semiconductors may be made using germanium with antimony or indium impurities. It should be observed that the resistivity of the pure germanium or silicon is reduced by the addition of donor or acceptor impurities.

p-n junction diode

A p-n junction is formed when p-type and n-type germanium or silicon are formed on a *single* continuous crystal. The normal random motion associated with the kinetic theory causes some of the holes $\oplus$ and the electrons $\ominus$ to diffuse

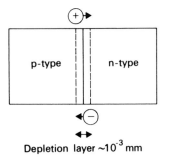

Depletion layer $\sim 10^{-3}$ mm

Figure 3.4   p-n junction without bias

across the junction causing a p.d. to be established, and this prevents further migration. The barrier thus formed at the junction is called the **depletion layer**. The p-type material now has a net negative charge, and the n-type a net positive charge.

Bias

A battery connected as in Figure 3.5(a) is called a **reverse bias**. This makes the n-type end more positive and the p-type end more negative, thereby effectively increasing the *width* of the depletion layer so that almost no current will flow.

An e.m.f. applied in the opposite direction, as shown in Figure 3.5(b), is called **forward bias**. The depletion layer is now overcome and holes from the p-type material may now flow through the n-type, with the electrons flowing in the opposite direction, completing the circuit. Currents in the mA range are observed.

Note: Conventional current direction is marked on all Figures in this section.

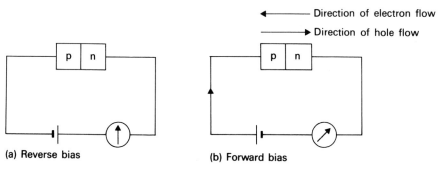

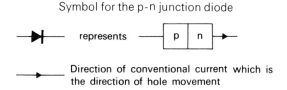

(a) Reverse bias   (b) Forward bias

Figure 3.5   p-n junction with bias

Thus this p-n junction acts as a **rectifier** as it preferentially allows the current to flow in the forward direction.

Symbol for the p-n junction diode

3:A.2 *Transistors*

A p-n-p transistor contains a thin section of n-type germanium or silicon between two p-type pieces on the same crystal. The n-p-n transistor has a thin section of p-type between two n-type pieces.

Symbols for transistors

p-n-p transistor   n-p-n transistor

b-base; c-collector; e-emitter (the arrow shows the direction of hole movement)

The **base** is the thin middle section of the p-n-p or n-p-n sandwich.

These transistors are called **bipolar** because there are mobile positive holes *and* mobile electrons in any single transistor.

Operation of the p-n-p transistor

For this transistor, emitter and collector refer to the 'emission' and 'collection' of positive holes. Notice that in the circuit shown in Figure 3.6 the emitter is common to both the base (input circuit) and the collector (output circuit), and therefore the transistor is said to be used in the **common emitter** condition.

If a bias is applied only across the transistor, that is the 5 V of Figure 3.6, there is almost no current as the depletion layers are not removed. When the base is given a small negative potential with respect to the emitter as shown by the 2 V battery, the base-emitter depletion layer is removed and holes from the emitter

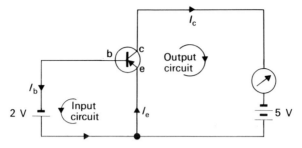

Figure 3.6 Operation of p-n-p transistor

pass into the base. A few combine with electrons of the base giving a small base-emitter current, but the majority pass through the thin base section into the p-type collector giving a larger collector-emitter current. If the base current ceases the depletion layer will re-form.

Note: The transistor only *conducts*, i.e. produces a current in the output circuit, when a suitable base potential is applied. Observe that the emitter/collector resistance is effectively a very *high* resistance when the transistor is *off*, or not conducting. However, this resistance is *low* when the transistor conducts and an emitter/collector current flows.

Thus the input current to the base *controls* the output circuit.

The transistor as an amplifier

The *characteristics* of a transistor are the sets of graphs showing the variation of collector or base current with either the p.d. between the base and emitter V_{be}, or the p.d. between the collector and emitter V_{ce}.

The **transfer characteristic** is the graph showing the variation of collector current I_c as the base current I_b is altered.

Experiment 3.1 *Current amplification by a transistor*

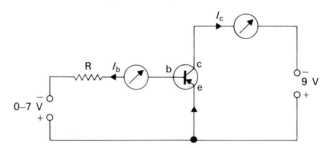

Figure 3.7 Circuit for obtaining characteristics

Using the circuit in Figure 3.7 the potential difference between the emitter and base V_{be} is varied and readings of collector current are recorded by the milliammeter. Either I_b may be calculated from the values of V_{be} or measured directly on a meter in series with R, (R $\sim$ 100 kΩ).

A graph of I_c against I_b (Figure 3.8) gives the transfer characteristic, and the gradient of this graph gives the **current amplification factor**.

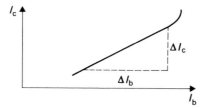

Figure 3.8 Transfer characteristic

The **current amplification factor** $= \dfrac{\Delta I_c}{\Delta I_b} = \dfrac{\text{change in collector current}}{\text{change in base current}}$

This is also termed the **current gain** or **forward current transfer ratio** of the transistor.

> As small changes in current in the input circuit produce large changes in current in the output circuit this device acts as a **current amplifier**.

As uniform current amplification is required, only the straight part of the graph is considered and the transistor is not operated outside this range. Typical values of the current amplification factor are around 50, with base currents about 100 μA and collector currents about 5 mA.

When a transistor heats up, the graph shown in Figure 3.8 will alter, causing the amplification to vary. Additional components are required to prevent this.

A practical transistor amplifier

The transistor in a common emitter condition with its base negative with respect to the emitter acts as a current amplifier.

The base may be maintained negative by:
(a) using a battery, as in Figure 3.6.
(b) using a potential divider method.

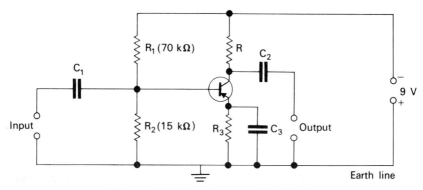

Figure 3.9 Transistor amplifier circuit

The bottom line of the circuit in Figure 3.9 is often earthed and is called the **earth line**. This implies that the battery maintains the upper end of the resistors R and R_1 at a negative potential of 9 V, with the positive terminal of the battery at earth potential. The potential divider consists essentially of the resistors R_1 and R_2 across the battery, the ratio of R_1 to R_2 determining the bias to the base. Notice that there is also a p.d. across R_3; the p.d. across R_2 minus the p.d. across R_3 is therefore the base bias.

With the base bias maintained either by a battery or by a potential divider, the input, usually an a.c. or varying current, is introduced through the capacitor as shown by C_1. The output is taken out via the capacitor C_2. The resistor R_3 and capacitor C_3 assist in preventing any large build up of current in the output circuit, which would cause the transistor to heat up and its characteristics to alter. For details on the function of the capacitors see page 127.

Note: The previous discussion may be applied to an n-p-n transistor providing all the battery connections are reversed so that the collector is positive with respect to the emitter. Compare the battery connections of Figure 3.10 below with those of Figure 3.7. With the n-p-n transistor the terms emitter and collector refer to 'emission' and 'collection' of electrons, but the arrow *still* points in the conventional current direction.

Experiment 3.2 *A transistor acts as a rapid switch*

An n-p-n transistor is used in this experiment.

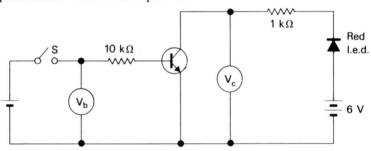

Figure 3.10 Transistor as switch

A light-emitting diode, an l.e.d., emits light when the current passes in the forward direction. When S is open the reading on V_b is zero and that on $V_c \sim 6$ V. The l.e.d. remains off, showing that no current flows in the collector/emitter circuit. V_c effectively measures the p.d. across the battery.

When S is closed V_b records the value of the base battery. The transistor conducts and the l.e.d. is lit immediately. The reading on V_c drops almost to zero since the emitter/collector p.d. is now negligible compared to the potential dropped across the l.e.d. and the 1 kΩ resistor.

A signal into the input circuit, S closed, can immediately control the output circuit.

This circuit illustrates *two* important uses of the transistor in modern electronics. First, information can be rapidly transmitted from one circuit to another by a transistor. Secondly, this circuit acts as a logical NOT gate. If V_b is zero or OFF then V_c has a reading or is ON, but if V_b has a reading or is ON then V_c is zero or OFF. The state of V_b is NOT the same as the state of V_c, assuming that only the *two states* of OFF and ON are considered. In practice six NOT, or inverting gates, are contained on one 14 pin chip, see Figure 3.13(c), as opposed to the individual construction of the above circuit.

Using the circuit of Figure 3.10, if the voltmeter V_c was placed across the 1 kΩ resistor then it would only show a reading when S was closed and V_b also indicating a reading. Now the transistor acts as a switch but without the negation function.

3:A.3 *Metal Oxide Semiconductor Transistors**

The metal oxide semiconductor MOS transistor is constructed from three materials, a **m**etal plate, an **o**xide layer and a **s**emiconductor region. The semiconductor region consists of a p-n-p or n-p-n sandwich. A section through a *p-channel enhancement mode* MOS, p-MOS transistor, showing the basic layout is given in Figure 3.11(a). The central n-type region has a *low* conductivity with only a very few mobile electrons. When a negative potential is applied to the metal plate, the **gate** electrode, positive holes will be attracted to the underside of the silicon oxide making the top part of this n-type material act as a p-type and provide a *channel* from the **source** to the **drain**, Figure 3.11(b). If the negative gate potential is increased in magnitude, the current from the source to the drain also will increase up to a maximum. Thus the gate potential *controls* the current through the transistor. Observe that the transistor acts as an amplifier, but as a *voltage* amplifier. (The symbol is shown in Figure 3.12(a).)

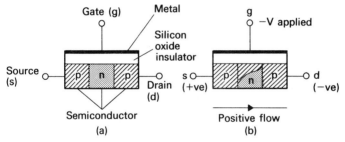

Figure 3.11 p-MOS transistor

The action of this transistor as a rapid switch is important. If the gate potential is zero no current flows from source to drain and the transistor is *off*. When a negative potential is applied to the gate the transistor is *on* and current flows from source to drain.

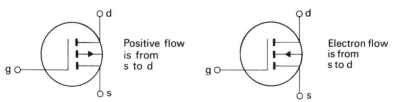

(a) p-channel enhancement mode MOS   (b) n-channel enhancement mode MOS

Figure 3.12 Symbols for MOS transistors

* Section 3:A.3 and 3:A.4 are for interest only.

The n-channel enhancement mode MOS transistor operates on the same principle, but now a *positive* potential is applied to the gate and *electrons* flow from source to drain. See Figure 3.12(b) for the symbol.

These MOS transistors are called **unipolar** since only *one* type of charge carrier is involved. In the p-MOS, holes are the charge carriers, and with n-MOS electrons carry the charge. (For interest it should be mentioned that there is another type of MOS transistor called the *depletion mode*, which has a slightly different construction and operation. Its symbol has a *solid* line linking source to drain). All these MOS transistors are termed **f**ield **e**ffect **t**ransistors FET since they operate under the action of the *electric field* applied to the gate.

Often a p-MOS is used in conjunction with a n-MOS in **c**omplementary MOS circuits, termed CMOS. These have great importance in microelectronics and microprocessor devices.

3:A.4 *The Integrated Circuit*

The transistor depends for its action on the extrinsic charge carriers. Although germanium has a larger number of intrinsic carriers than silicon, the earlier popularity of germanium is being displaced by silicon leading to the 'silicon chip'.

The silicon planar transistor can be manufactured, as just one of about a thousand similar transistors, from a single slice of n-type (or p-type) silicon, termed the substrate, which is then suitably doped to give the p and n regions. Each transistor is *very* small hence the term **silicon chip**, Figures 3.13(a) and (b).

Instead of making individual transistors, resistors, capacitors, and diodes, the entire circuit can be made on one small piece of silicon by doping various parts and making appropriate connections. The circuit is then termed an **integrated**

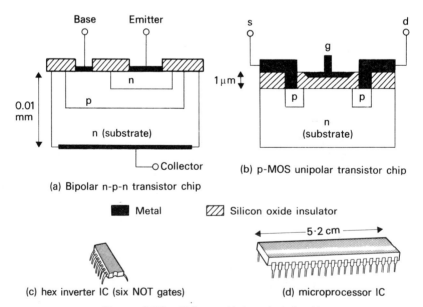

(a) Bipolar n-p-n transistor chip

(b) p-MOS unipolar transistor chip

(c) hex inverter IC (six NOT gates)

(d) microprocessor IC

Figure 3.13 Chips and integrated circuits

circuit IC. A *microprocessor* is an integrated circuit which contains a unit for performing arithmetic and logical operations, a *control* unit for supervising the execution of instructions, and special storage registers. A *microcomputer* contains a microprocessor, together with memory integrated circuits to store information, supporting circuits, and input and output devices. For example the input device could be a typewriter keyboard and the output device a TV screen. This very simplified outline is given to indicate the importance of the integrated circuit.

Note: It should be remembered, throughout this chapter, that where values of components are given on diagrams these are only typical *not* essential values.

Problems

3.1 An n-type semiconductor crystal contains a small amount of an acceptor, antimony for example, so that the crystal has an excess of electrons. Is this statement correct?

3.2 What is (*a*) the transfer characteristic of a transistor, (*b*) the current amplification factor?

3.3 Comment on the relative values of (i) the high resistance voltmeters V_1, V_2 and V_3 and (ii) the ammeters A_1 and A_2 when (a) S is open (b) S is closed.

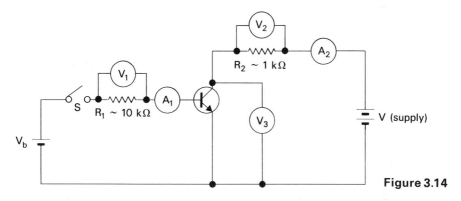

Figure 3.14

B: Alternating Current

3:B.1 Sinusoidal Alternating Current

A potential difference maintaining a direct current, d.c., can have a steady value V in one direction. With an alternating current, a.c., the *direction* of the supply *varies* with time and hence the direction of the current will change.

Note: Because the following discussion is concerned with a comparison between a.c. and d.c. the word voltage will be used for both p.d. and e.m.f. However, the distinction between the e.m.f. of an a.c. supply and the p.d. between two points in a circuit should be borne in mind.

Sinusoidal voltage

With the common form of alternating current, the supply e.m.f. varies continuously from a maximum in one direction to a maximum in the other. This alternating variation is sinusoidal with time as shown in Figure 3.15.

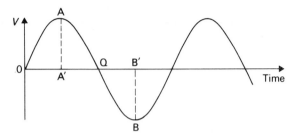

Figure 3.15 Variation of V with time (sine curve)

The maximum value in either the positive or negative direction, AA′ or BB′, is termed the **peak voltage**, V_m. The *peak-to-peak* value is twice V_m.

The energy or power from an alternating source will be the average over time of the instantaneous energy or power. For a peak value of V_m the power at time A′ through a resistor of resistance R will be $\dfrac{V_m^2}{R}$ but at time Q it will be zero.

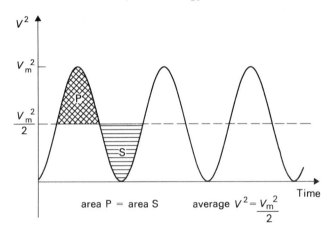

Figure 3.16 Graph of V^2 and time

For a given resistor the power $\propto V^2$. From the graph shown in Figure 3.16 it can be seen that the average value of V^2 with time is $\frac{1}{2}V_m^2$. Thus the **effective voltage** is $\dfrac{V_m}{\sqrt{2}}$. A d.c. supply of *constant* voltage $\dfrac{V_m}{\sqrt{2}}$ will give the same power as a sinusoidal a.c. supply of peak value V_m. Hence as expected, the equivalent d.c. value is *less* than the peak a.c. value.

Note: The **effective voltage**, termed the **root mean square** (r.m.s.) voltage of an alternating source, is defined as that direct voltage which gives the *same* power as the alternating source.

The **effective**, or r.m.s., **current** is defined in a similar way: as that direct current which gives the same power as the alternating current.

$$V_{\text{r.m.s.}} = \frac{\text{peak voltage}}{\sqrt{2}} = \frac{V_{\text{m}}}{\sqrt{2}} \qquad I_{\text{r.m.s.}} = \frac{\text{peak current}}{\sqrt{2}} = \frac{I_{\text{m}}}{\sqrt{2}}$$

Hence the average power dissipated $= V_{\text{r.m.s.}} \times I_{\text{r.m.s.}} = \dfrac{V_{\text{m}} I_{\text{m}}}{2}$.

An alternating voltage is usually *quoted* as an r.m.s. value. For example the 240 V mains have an r.m.s. of 240 V.

The **frequency** f of an alternating source is the number of complete cycles which occur in one second and is measured in **hertz**, Hz. As with all oscillations the **period** T is the time for one complete cycle and $f = \dfrac{1}{T}$.

Note: Unless otherwise stated, an alternating supply is assumed to be sinusoidal with time.

Example: A $10\,\Omega$ resistor produces heat at the rate of 22.5 joules per second. State the p.d. across, and the current through, the resistor if (*a*) the supply is d.c. and (*b*) the supply is a.c. (Give peak values.)

(*a*) Using power $= I^2 R$ then $22.5 = I^2 \times 10$ and $I = 1.5\,\text{A}$
    and $V = IR$ then $V = 1.5 \times 10$ and $V = 15\,\text{V}$
   For a d.c. supply the p.d. is 15 V and the current 1.5 A.

(*b*) For an a.c. supply the effective, or r.m.s., values will be the same.
   Thus $V_{\text{r.m.s.}} = 15\,\text{V}$ and $I_{\text{r.m.s.}} = 1.5\,\text{A}$.

    Giving peak values of $V_{\text{m}} = 15 \times \sqrt{2} = 21.2\,\text{V}$

       and $I_{\text{m}} = 1.5 \times \sqrt{2} = 2.1\,\text{A}$

   Both the current and p.d. vary continuously from these values to zero.

In some instances, such as in the insulation of electric cables, the peak value of the current may be important.

Experiment 3.3 *To determine the effective voltage of an a.c. source*

The brightness of the lamp in Figure 3.17 is observed when the a.c. source is connected by joining M to S_1. A cathode ray oscilloscope (see Section 3:B.3) may be used to measure peak voltage and this value should be noted. M is then joined to S_2 and the resistance altered until the lamp has the same brightness. This can be checked by flipping the switch between S_1 and S_2, or more accurately by using a **light meter** placed next to the lamp. The direct voltage then provides the same power as the alternating voltage.

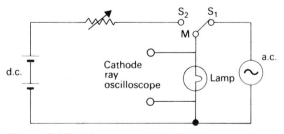

Figure 3.17 Measurement of effective a.c. voltage

This experiment shows that for the same power, the direct voltage is *less* than the peak value of the alternating voltage, and:

$$V_{d.c.} = \frac{V_{peak\,(a.c.)}}{\sqrt{2}} = \text{effective voltage (a.c.)}$$

3:B.2 *Measurements in Alternating Current Circuits*

Measurement of current

(1) A **moving-coil galvanometer** may not be used directly to measure an alternating current as the coil will turn in the opposite direction when the current is reversed. However, if a rectifier (see Section 2:C) is used in conjunction with the galvanometer this instrument may be used to measure alternating current.

(2) A **moving-iron ammeter** is usually less sensitive and less accurate than the moving-coil galvanometer, but it has the advantage of measuring alternating current directly without rectifying circuits.

One design of this meter, the *attraction type*, has a piece of soft iron suspended near a fixed coil through which the current to be measured is passed. The iron becomes a temporary magnet, due to the coil's magnetic field, and is attracted towards the coil. When the current changes direction the magnetic field inside the coil reverses, together with the magnetic polarity of the iron. Hence the force on the iron remains attractive.

Another design, the *repulsion type*, has two pieces of soft iron placed side by side inside a coil, one of them PQ being fixed. When the a.c. is passed through the coil the pieces of iron become temporary magnets with the *same* polarity and repulsion of the movable piece XY occurs, Figure 3.18.

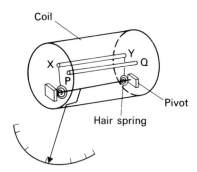

Figure 3.18   Repulsion type moving iron meter

In both designs the magnitude of the repulsion or attraction will depend on the magnitude of the current and *not* on the alternating nature of the current.

Note: This ammeter has a non-linear scale.

Measurement of frequency and potential difference

(1) A **c.r.o.** (see below) may be used to measure both frequency and alternating p.d. Remember, the c.r.o. displays *peak* values V_m, the r.m.s. values must be

calculated from $\dfrac{V_m}{\sqrt{2}}$. The c.r.o. has the advantage of having a very *high impedance* (see Section 4:C.1). Also notice that the inertia of the electron beam is negligible.

(2) The **moving-coil galvanometer** with a rectifier may also be used to measure alternating potential difference.

3:B.3 *Cathode Ray Oscilloscope c.r.o.*

The cathode ray oscilloscope is an evacuated glass tube containing an electron gun and deflecting plates (Figure 3.19).

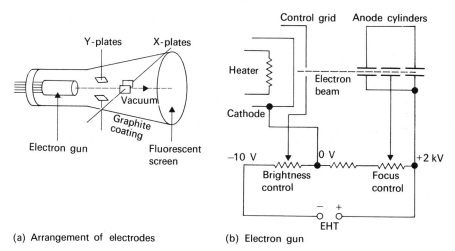

(a) Arrangement of electrodes   (b) Electron gun

Figure 3.19 C.r.o.

A beam of electrons emitted by the cathode is collimated by the cylindrical focusing anode. The beam may be deflected vertically by applying a p.d. between the Y-plates or horizontally by applying a p.d. between the X-plates, before impinging on a fluorescent screen. A **time base** is frequently applied to the X-plates so that the spot is swept across the screen at a uniform speed with a very fast 'fly back' return. This time base has a saw-tooth wave form. A simple time base circuit is discussed in Section 4:A.4. The number of sweeps per second is called the **time base frequency**.

The wave form or p.d. to be examined is applied to the Y-plates via an amplifier. The **gain** control of the amplifier may therefore determine the **amplitude** of the Y-plate signal.

Controls of the c.r.o.

The chief controls are described briefly below.

(1) **Focus**: to produce a collimated beam, see Figure 3.19(b)

(2) **Brilliance**: the control grid determines the number of electrons emitted, see Figure 3.19(b). Also, the power is turned on and off by this knob.

(3) **Y gain** or **Volts/cm**: to determine the amplitude of the vertical deflection.

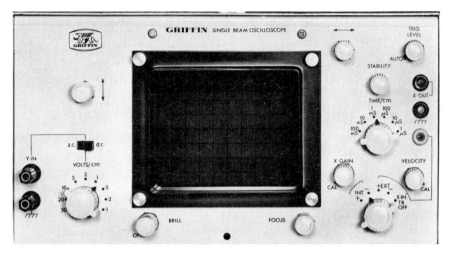

Figure 3.20   c.r.o. controls (*Griffin and George,* single beam oscilloscope)

(4) **a.c./d.c.**: switch for a high or low frequency input.

(5) ↕ **Y shift**: to move the display vertically up and down.

(6) **Y** and adjacent **earth sockets**: the input terminals for the application of a signal to the Y-plates.

(7) ←→ **X shift**: to shift the display horizontally.

(8) **X gain**: to expand the display horizontally.

(9) **Time/cm** or **T/B**: time-base switch to select the time-base frequency.

(10) **Velocity** control: for fine adjustment of the time base in the chosen range.

(11) **Trigger selector** switch: with positions for the internal INT $\pm$, or an external EXT $\pm$ time-base signal to be connected to the X-plates. The time-base off position, X IN/TB OFF, allows an external signal to be applied to the X-plates. (On some instruments there is not a separate selector switch but the T/B switch (10) has an OFF position.)

(12) **EXT/X** and adjacent **earth sockets**: input terminals for an external signal to the X-plates. (These sockets are at the rear of some instruments.)

(13) The **screen**, with a square grid.

The controls (7) and (8) are omitted on certain models.

Applications of the c.r.o.

(1) Potential difference and e.m.f. measurements, with the advantage that a c.r.o. takes almost no current, having a very high impedance.

(2) Wave form display, for example to examine smoothing circuits. (See Section 4:A.4).

(3) Frequency display and measurement.

(4) Time measurement.

Lissajous' figures

These result from the compounding of two wave motions at right angles to each other. The time base of a c.r.o. is switched to the 'off' position and a signal generator is connected to the X-plates. Another alternating input signal is connected to the Y-plates. The spot on the screen is now oscillating in both the X and Y directions. The frequency of the signal generator is adjusted to obtain stationary patterns known as the Lissajous' figures.

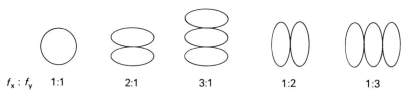

$f_x : f_y$ 1:1   2:1   3:1   1:2   1:3

Figure 3.21   Lissajous' figures

From the patterns an unknown frequency signal to the Y-plates may be determined.

Note: The difference in the patterns $f_x : f_y = 1:2$ and $f_x : f_y = 2:1$.

Problems

3.4   An a.c. moving coil galvanometer registers 3.1 A. What is the peak value of the current, and how many times does the current fall to zero in one second, if the frequency is 50 Hz?

3.5   A c.r.o. displays a sinusoidal p.d. of peak value 30 V. The time taken for one complete cycle is 25 ms. State the frequency of the supply and the heat produced per second, if this p.d. is applied across a 6 Ω resistor.

3.6   An a.c. signal is applied to the Y plates of the c.r.o. and the trace obtained is shown in Figure 3.22. Assume that each square in the figure is 1 cm by 1 cm. The controls were set as follows: time base at 0.5 ms cm^{-1} and Y gain at 50 mV cm^{-1}. Determine:
(*a*) the time for one cycle and the frequency of the signal.
(*b*) $V_{r.m.s.}$ of the applied p.d.

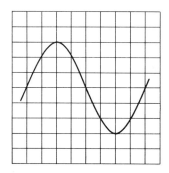

Figure 3.22

3.7   The input to the X-plates is 80 Hz and an unknown frequency f is applied to the Y-plates. The pattern in Figure 3.23 is obtained. What is the frequency of f?

   Figure 3.23

3.8   The contact K is vibrated between N and M. P is connected to the Y earth input socket and Q to the other Y input socket.

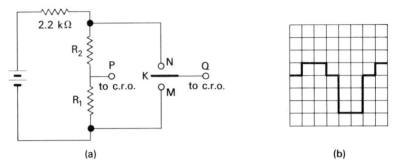

| (a) | (b) |

Figure 3.24

The trace shown in Figure 3.24(b) is obtained. Assume the squares to be 1 cm squares. The c.r.o. controls were set at $2\,\text{V}\,\text{cm}^{-1}$ and $0.1\,\text{ms}\,\text{cm}^{-1}$.
(a) State the time that K is in contact with N.
(b) State the frequency of the vibration of K.
(c) If $R_1 = 3.3\,\text{k}\Omega$ what is the value of R_2?
(d) Determine the current through the $2.2\,\text{k}\Omega$ resistor and the terminal potential difference of the source.

C: Rectifiers

A **rectifier** is a circuit component for producing a unidirectional current from an a.c. supply.

(1) The diode valve

The diode is a valve with two electrodes, cathode, and anode, enclosed in an evacuated glass envelope. The heated cathode emits electrons which are attracted to the anode, only when the anode is positive with respect to the cathode. A current is thus allowed in one direction only, that is, from anode to cathode, with the electrons travelling from cathode to anode.

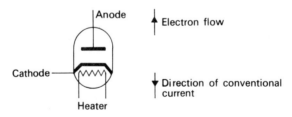

Figure 3.25   Diode

Note: The emission of electrons from a heated metal is termed **thermionic emission**.

(2) **The p-n junction** diode is the most widely used rectifier. It is cheaper, smaller and less easily damaged, since there are no glass envelopes to break. Also no heater circuits are required.

Note: Semiconductor devices, such as the p-n junction or transistor, are often termed 'solid-state' devices (in contrast to the valves with their glass envelopes).

Half-wave rectification

For either (1) or (2) above the symbol ➤➤ will be used in circuit diagrams to indicate a rectifier.

With the circuit in Figure 3.26 the rectifier prevents the passage of the negative half-cycles and a pulsating unidirectional current is produced.

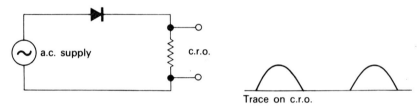

Figure 3.26 Simple rectifying circuit

Full-wave rectification

(1) *Use of two diodes or p-n junctions*

The conventional direction of the current will be from B to A through the resistor. When X is positive, current flows through D_1, then from B to A back to

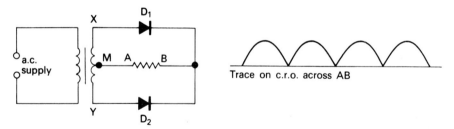

Figure 3.27 Rectifying circuit with centre tapped transformer

M. When Y is positive, in the next half cycle, current flows through D_2, then from B to A back to M. Here both half cycles of the a.c. supply are utilized.

(2) *Bridge circuits*

During one half cycle the conventional current will flow PRA→BSQ and during the other half cycle the current direction is QRA→BSP. Notice that in both half cycles the conventional current is A→B. This has the advantage that the total potential is applied across the resistor in each half cycle whereas in the full-wave rectifier (1) only half the potential difference XM or MY is effective. Solid state rectifiers are usually employed to avoid separate heating circuits.

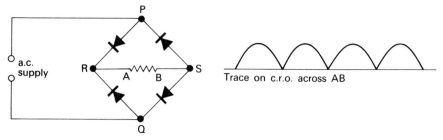

Figure 3.28   Rectifying bridge circuit

A bridge circuit using p-n junction diodes is used in the adaptor to convert a moving coil galvanometer for use in a.c. circuits.

Note: Rectifiers can produce a varying unidirectional current from an a.c. supply. To produce a steady direct current these variations must be reduced. This is called *smoothing*. Capacitors and inductors are used to achieve smoothing, (see pages 112 and 122).

Problems
3.9   Why is the diode enclosed in an evacuated glass envelope?
3.10   Give two advantages of the solid-state rectifier over the diode.

4 Capacitors and Inductors as Circuit Elements

A: Capacitance

4:A.1 The Electroscope

Since a *conducting* solid has a large number of mobile electrons, a charge placed on it will be able to migrate throughout the bulk of the material. In a conducting solution, the charge carriers are the positive and negative ions.

A *semiconductor* has a smaller number of mobile charge carriers which may be either positive or negative, that is electrons or positive holes (see Section 3:A.1).

A non-conducting substance, an *insulator*, has effectively no mobile charge carriers and therefore a charge placed on an insulator will tend to remain in the region in which it is placed.

$$\text{Conductors: low resistivity} \sim 10^{-7}\,\Omega\,\text{m}$$
$$\text{Semiconductors: resistivity} \sim 10^{2}\ \ \Omega\,\text{m}$$
$$\text{Insulators: high resistivity} \sim 10^{10}\ \ \Omega\,\text{m}$$

The electroscope

This instrument has a conducting plate P connected by a central spine to a gold leaf L. The spine is separated from the box, which is also a conductor, by an insulating stopper S. When the plate P is given a charge this will spread to both the end of the spine and the leaf, which will repel each other, causing the divergence of the leaf. If the box were given a charge of the same sign, its repulsion would oppose the divergence of the leaf. In most arrangements the box is earthed to obviate this effect, and the divergence of the leaf is an indication of the **potential difference** between the electroscope plate and earth.

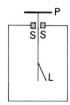

Figure 4.1 A simple electroscope

Charging by sharing and induction

When a negatively charged conductor is touched on to the plate P of the electroscope some electrons will pass across to the plate and down to the leaf causing divergence. The electroscope has been charged by **sharing**.

A polythene rod may be given a negative charge by rubbing it with a woollen cloth. If instead of touching the plate, the rod is held *near* the plate as in Figure

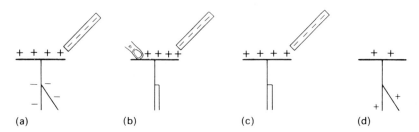

(a)   (b)   (c)   (d)

Figure 4.2 Charging by inductor

4.2(a), the plate is touched by the finger (b), the finger removed (c), and then the rod removed (d), a positive charge is induced on the plate P, the electrons being repelled away down the finger to earth. This is called charging by **induction**.

Note: The charge obtained by induction has the opposite sign to the charging body. In sharing, charge of the same sign is obtained. Also a larger charge may be transferred by induction, usually equal to the charge on the charging body, whereas only a small part of the charge is transferred in sharing.

Earth

Symbol

Earth is some large conducting body which may supply or receive charges without a noticeable change in its own state of charge.

4:A.2 *Capacitance and Capacitors*

A simple **capacitor** is usually composed of two metal conducting plates separated by air or an insulator.

Symbol for a capacitor

Its function is to *store* charge.

(*a*) If a negative charge is put on one plate B in Figure 4.3(a), then electrons will be repelled off the other plate A leaving it with a positive charge.

(*b*) If the capacitor is connected to a battery, as in Figure 4.3(b), it will also store charge. The amount of charge stored now depends on the e.m.f. of the battery as well as the design of the capacitor. The charge stored by a capacitor refers to the charge on *either* plate.

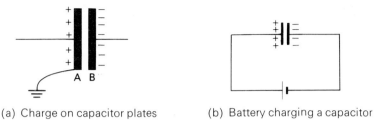

(a) Charge on capacitor plates   (b) Battery charging a capacitor

Figure 4.3

When the p.d. across the plates V is increased, the charge Q on the capacitor will also increase. It can be shown for a given capacitor that $Q \propto V$.

Experiment 4.1   *Qualitative demonstration that Q ∝ V*

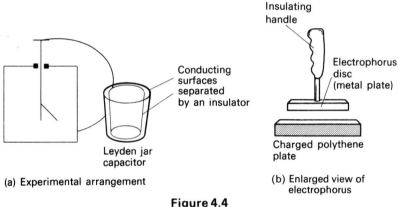

(a) Experimental arrangement

(b) Enlarged view of electrophorus

Figure 4.4

The electrophorus disc is charged by *induction* using the polythene plate. Each time the disc is charged by the plate it carries the *same* quantity of charge. The charged disc is placed inside the Leyden jar capacitor and the charge is transferred to the inside of the jar. It can be assumed that *all* the charge on the disc is transferred. Hence if the operation is repeated a number of times N, then N units of charge are transferred. The electroscope measures p.d., thus the deflection of the leaf indicates the p.d. across the jar.

It is observed that the deflection increases with the number of strokes indicating that:

$$V \propto Q$$

Note: The constant of variation is the capacitance C giving $Q = CV$.

An important quantitative experiment showing that $Q \propto V$ is illustrated by Problem 4.2 page 113.

Capacitance

Unit: farad (F), scalar.

The capacitance of a capacitor is a measure of its ability to store charge.

The capacitance of a capacitor is defined as the charge required to raise the potential difference between the plates to one volt.

$$\text{Capacitance}_{\underset{F}{|}} = \frac{\overset{C}{|}\text{charge}}{\underset{|}{\text{potential difference}}_{\underset{V}{|}}}$$

$$C = \frac{Q}{V}$$

Definition of the farad

If one coulomb of charge is needed to provide a potential difference of one volt between the plates of a capacitor then the capacitance is one farad.

The farad is too large a unit for practical capacitors and smaller units are usually required:

$$\text{microfarad } \mu\text{F } 10^{-6}\text{ F}$$
$$\text{picofarad pF } 10^{-12}\text{ F}$$

Example: A capacitor stores 8×10^{-4} C of charge when the potential difference between the plates is 100 V. What is the capacitance?

$$C = \frac{Q}{V} = \frac{8 \times 10^{-4}}{100} = 8\,\mu\text{F}$$

Notice in 8×10^{-4} C, the C stands for the unit coulomb but in formulae or electrical circuits the symbol C denotes capacitance.

Experiment 4.2 *To show that a capacitor stores charge*

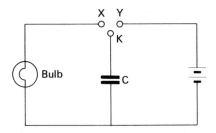

Figure 4.5 A capacitor can store charge

Using the circuit shown in Figure 4.5, K is joined to Y to charge the capacitor. Then K is joined to X to discharge the capacitor and the bulb lights up for a short time. Since the bulb lights up, a current must have passed. Thus the capacitor stores *charge* and *energy*. The energy stored by the capacitor is provided by the battery.

Energy stored in a capacitor

A capacitor can store charge of opposite sign on its plates. If the plates are connected by a wire a current will flow and the stored energy will be dissipated as heat. A resistor can produce heat from electrical energy when a current flows, but if the current ceases no more heat is obtained. A resistor cannot store electrical energy, but a capacitor *can* store energy.

Note: Energy must be provided to place charge on the capacitor plates. Also, energy is converted when a charged capacitor discharges.

To determine the energy stored by a capacitor, consider the work done to place a charge Q on the uncharged capacitor plate. The initial p.d. between the plates is zero and the final p.d. is V. Hence the average p.d. is $\frac{1}{2}V$.

$$\text{Work done} = \text{charge} \times \text{p.d.}$$

hence $\qquad$ work done to charge capacitor $= Q \times$ average p.d.

$$\Rightarrow \text{energy stored by the capacitor} = \tfrac{1}{2}QV$$

Note: For a given capacitor the p.d. V between the plates varies directly as the charge Q stored, $(CV = Q \Rightarrow V \propto Q)$. Thus from the graph (Figure 4.6),

$$\text{Energy} = \text{charge} \times \text{p.d. at each instant of time}$$
$$= \text{area under the graph}$$
$$= \tfrac{1}{2}Q_0 V_0$$

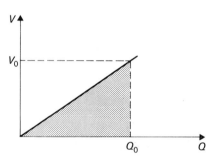

Figure 4.6

Using $C = \dfrac{Q}{V}$

$$\boxed{\text{Energy stored by a capacitor} = \tfrac{1}{2}Q_0 V_0 = \tfrac{1}{2}CV_0^2 = \tfrac{1}{2}\frac{Q_0^2}{C}}$$

where Q_0 is the charge stored and V_0 the final p.d. between the plates.

Example: A cell of e.m.f. 4 V is used to charge a 500 μF capacitor. Determine the charge on each plate and the energy stored in the capacitor.

Using $\qquad\qquad C = \dfrac{Q}{V}$ then $Q = 4 \times 500 \times 10^{-6} = 2\,\text{mC}$

Using $\qquad\qquad$ energy $= \tfrac{1}{2}CV^2$

gives $\qquad\qquad$ energy stored $= 4 \times 10^{-3}\,\text{J}$

Factors affecting the capacitance of a capacitor

The ability of a capacitor to store charge, that is its capacitance, depends on its shape and size.

Experiment 4.3   *Parallel plate capacitor*

One plate of the capacitor is connected to the disc of the electroscope and is given a positive charge by induction. The other plate of the capacitor is earthed and negative charges are drawn up on to it by the attraction of the positive plate. The divergence of the leaf indicates the potential difference between the positive plate and earth.

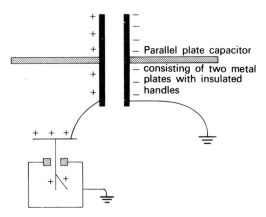

Figure 4.7   Capacitance of a capacitor

During the experiment the position of the earthed plate is altered, but care is taken not to touch the positive plate so that the charge Q on it remains constant. Any change in the divergence of the leaves indicates a change in p.d. V and since $C = \dfrac{Q}{V}$ this can only be caused by a change in capacitance C, Q being constant.

Note: An *increase* in divergence means an increase in p.d. V which in turn means a *decrease* in capacitance C.

The following table is completed by experiment:

Alteration	*Leaf divergence*	*Capacitance*
Move plates together	decrease	increase
Move plates apart	increase	decrease
Decrease area of overlap	increase	decrease
Increase area of overlap	decrease	increase
Place glass between plates	decrease	increase

This experiment shows that the capacitance is increased if the area of overlap A of the plates is increased. Larger plates may store more charge, but if the plates are not opposite to each other less charge will be attracted on the other plate.

When the separation d of the plates is small the attraction on one plate of charges of opposite sign on the other plate is increased, so helping to overcome the repulsion of neighbouring like charges on that plate. Again more charge may be stored.

The capacitance is also increased if material of a larger **dielectric constant** ε_r is placed between the plates. (ε_r is also termed the **relative permittivity**.)

It may be shown that:

$$C \propto A, C \propto \frac{1}{d} \text{ and } C \propto \text{ relative permittivity, } \varepsilon_r$$

Types of dielectric and working voltage

The insulating material between the plates of a capacitor is called the **dielectric**. The **dielectric constant** of a material which fills the space between the plates of a capacitor is the factor by which it will increase the value of the capacitance compared with a vacuum between the plates.

Example: The dielectric constant of air is almost unity, the same as a vacuum. The capacitance of a capacitor is measured with air between the plates then observed to double when waxed paper is placed between the plates. The dielectric constant of waxed paper is therefore two.

Examples of values of dielectric constants

Material	Dielectric constant
Air (vacuum)	1
Waxed paper	2
Mica	7
Ceramics	6–4000

It is important for the dielectric to *remain* an insulator even if the p.d. across the capacitor increases. The **working voltage**, wV, is the maximum p.d. which can be applied across the capacitor without causing the insulation of the dielectric to alter. If a larger p.d. is applied, a spark might pass between the plates. The p.d. which causes the capacitor to pass a current is called the **breakdown voltage**.

Practical types of capacitor

(1) Variable air capacitor

This consists of a number of parallel, semicircular-shaped plates on a central spindle connected alternately together, (Figure 4.8). The dielectric is air. This capacitor may be used as a tuning capacitor in radios. Typical values are around $3 \times 10^{-4} \, \mu F$.

(2) Electrolytic capacitor

Symbol

The aluminium electrolytic capacitor has a thin layer of aluminium oxide, the dielectric, deposited on a sheet of aluminium foil, the positive plate. The negative

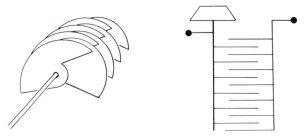

Figure 4.8 Variable air capacitor

plate is a piece of paper impregnated with the electrolyte. Another sheet of aluminium foil is placed next to the electrolyte paper to make the external connections. A sheet of paper insulation is added before the whole arrangement is rolled into a cylinder to reduce the overall size.

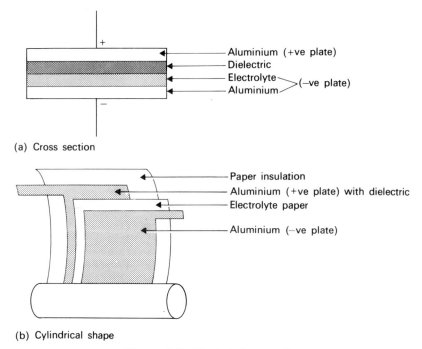

(a) Cross section

(b) Cylindrical shape

Figure 4.9 Electrolytic capacitor

These capacitors must not be charged in the reverse direction or the oxide layer will decompose, their positive and negative terminals being marked. Notice the symbol for these capacitors. The dielectric layer is extremely thin hence high values of capacitance are possible without the capacitor being unduly large. Values range from 0.47 μF up to 4700 μF with a range of working voltages.

The *tantalum electrolytic capacitor*, which is usually smaller in size, has tantalum electrodes with tantalum pentoxide as the dielectric.

(3) *Paper capacitor*

Two long sheets of thin metal foil are placed in contact with two long sheets of paper impregnated with paraffin wax and rolled into a cylinder. These capacitors tend to be quite large but with high working voltages, for example, a $2\,\mu F$ capacitor with a wV of 600 V.

Capacitors in parallel*

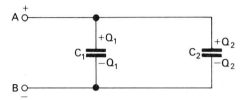

Figure 4.10   Capacitors in parallel

Let C be the total capacitance between A and B.
   If V is the p.d. between A and B then

$$\text{p.d. across } C_1 = \text{p.d. across } C_2 = V$$

The total charge Q is shared between the two capacitors,

$$Q = Q_1 + Q_2$$

$$CV = C_1 V + C_2 V \qquad \text{using } C = \frac{Q}{V}$$

$$C = C_1 + C_2$$

This result may be extended to any number of capacitors.

Note: The total capacitance of capacitors in parallel is the *sum* of their separate capacitances.

Capacitors in series*

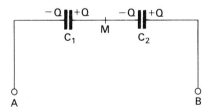

Figure 4.11   Capacitors in series

Let C be the capacitance of the single capacitor which may replace capacitors C_1 and C_2.
   When charging the capacitors a charge of $-Q$ is induced on the left-hand

---

*The formulae for capacitors in parallel and series are given for interest only.

plate of C_1 and $+Q$ on the right-hand plate of C_2. Thus charges of $+Q$ and $-Q$ will be induced on the inner plates of C_1 and C_2 respectively. The *net* charge on the inner plates, which are connected, remains at zero.

If V is the p.d. between A and B, V_1 is the p.d. AM across C_1 and V_2 the p.d. MB across C_2 then

$$V = V_1 + V_2$$

$$\frac{Q}{C} = \frac{Q}{C_1} + \frac{Q}{C_2} \qquad \text{using } C = \frac{Q}{V}$$

$$\frac{1}{C} = \frac{1}{C_1} + \frac{1}{C_2}$$

Note: The total capacitance of capacitors in series is always *less* than the capacitance of any one of them.

4:A.3   *Charging and Discharging a Capacitor*

A capacitor takes *time* to charge or discharge.

Experiment 4.4   *Variation of p.d. across the capacitor*

A capacitor is connected to a d.c. supply through a resistor. Suitable values for the components are shown in Figure 4.12.

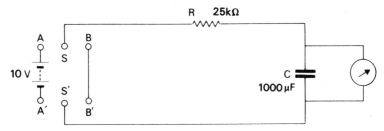

Figure 4.12   Charging and discharging a capacitor

Note: Where values of components are given it should be understood that these values are only typical *not* essential.

After joining A to S and A′ to S′ the potential difference across the capacitor is recorded on the voltmeter as the capacitor C is charging up. This is repeated with other values of R and C.

The supply is removed and the capacitor short circuited by joining B to S and B′ to S′. The potential difference across C is again recorded during discharging.

When charging, a capacitor *takes time* to build up its charge to a maximum potential difference. Also when discharging, the charge *takes time* to flow off the capacitor plates reducing the potential difference to zero.

As a larger capacitor is able to store more charge it takes longer to build up to its maximum potential difference, and longer to lose all its charge when discharging. A larger resistance only allows a smaller current to flow, hence more time is required for the charge to flow on and off the plates. Thus it is the product RC which determines the *time taken* for the capacitor to charge up to its maximum p.d. For interest it may be mentioned that RC is called the **time**

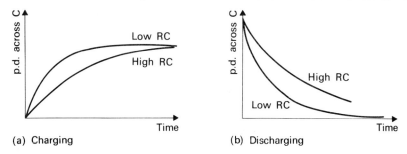

Figure 4.13 Variation of p.d. with time

constant. (The unit of RC is the second; ohm $= \dfrac{\text{volt}}{\text{ampere}}$, farad $= \dfrac{\text{coulomb}}{\text{volt}}$ giving ohm $\times$ farad $=$ second.)

Experiment 4.5 *Variation of current in a circuit*

A capacitor is again connected to a d.c. supply through a resistor. Typical values are given in Figure 4.14.

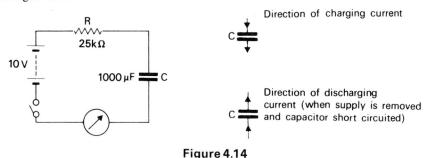

Figure 4.14

The current I flowing in the circuit as the capacitor charges up, after closing the switch, is recorded by the ammeter. This is repeated with different values of R and C.

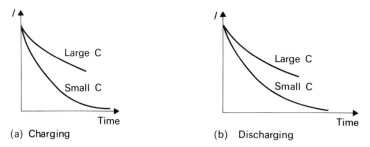

Figure 4.15 Variation of current with time

When the switch is closed the full potential difference of the supply is across the circuit so that the maximum current will flow. As charge flows on to the capacitor plates the rate of flow of extra charge (the current) will decrease until the capacitor is fully charged, when the current will be zero.

If a larger capacitor is used, more charge must be put on its plates so that the current build up will be longer. As in Experiment 4.4, a larger resistance merely reduces the *size* of the current.

Similar graphs are obtained during discharging except that the current is now in the *reverse* direction as the charge flows *off* the capacitor plates.

Note: Remember that to discharge the capacitor the supply is removed and a short circuit, or a resistor, is put in its place. If the switch of Figure 4.14 is merely opened, the capacitor would *remain* charged.

4:A.4 *Uses of a Capacitor*

Use of a capacitor to smooth rectified a.c.

Experiment 4.6   *The smoothing capacitor*

In this experiment a diode and various capacitors are used to endeavour to obtain a d.c. output from an a.c. supply. Again typical values are shown in Figure 4.16.

The Y-plates of a cathode ray oscilloscope are connected across the 1 kΩ resistor. The traces obtained are shown on the right of Figure 4.16.

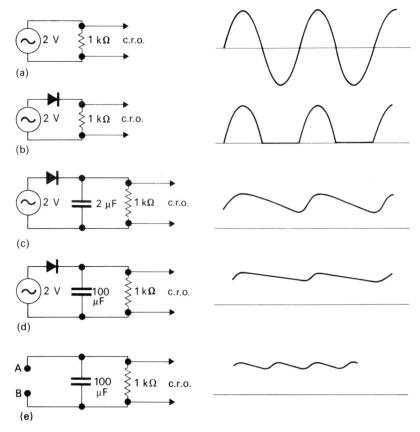

Figure 4.16   Use of a capacitor in smoothing

In (a) an alternating voltage trace is obtained.

In (b) the diode allows current to pass only in one direction and therefore the negative half cycles are eliminated.

In (c) the capacitor charges up each half cycle, but as it takes time to discharge, the potential difference will not have time to fall to zero before the next charging half cycle.

In (d) a larger capacitor is used which has a slower rate of discharging and so will 'hold up' the p.d. and give better 'smoothing'.

This is an example of smoothed half-wave rectification.

In (e) full wave rectified a.c. (from a circuit such as Figure 3.28), is applied at AB across the capacitor. Since there are two charging half cycles, better smoothing is achieved.

Use of a capacitor to generate an approximate saw-tooth wave form

An approximate saw-tooth wave may be obtained with the circuit shown in Figure 4.17. The capacitor C charges up to the striking potential of the neon tube which then ionizes and quickly discharges the capacitor. This process is then

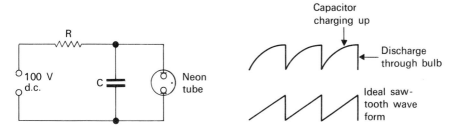

Figure 4.17   Generation of approximate saw-tooth wave form

repeated. The values of C and R determine the charging-up time. To increase the frequency of the waveform the value of R or C must be decreased. This circuit could be used to provide the time base of a c.r.o., the charging-up process giving the sweep across the screen and the discharge producing the rapid 'fly back'. In practice a more complicated circuit is used.

Another use of a capacitor is to *block* d.c. but transmit a.c. signals, see page 126.

Problems
4.1 A cell of e.m.f. 4 V and negligible internal resistance is connected in series with a $1000 \mu F$ capacitor, a $1 k\Omega$ resistor R, an ammeter, and a switch S. When S is closed state:

(*a*) the initial reading on the ammeter,

(*b*) the maximum and minimum p.d. across the resistor,

(*c*) the maximum and minimum p.d. across the capacitor,

(*d*) the final charge and energy stored by the capacitor.

4.2 Use the circuit indicated in Figure 4.18. The variable resistor R was adjusted to its maximum value.

The supply was switched on and the resistor R adjusted to keep the milliammeter reading constant at 0.3 mA. The following readings of V were taken every 3 s:

V (volt)   0   0.8   1.6   2.4   3.0   4.0   4.8   5.4   6.2   7.0   7.8   8.6

(*a*) Plot a graph of V against time.

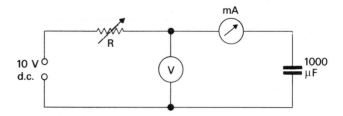

Figure 4.18 Quantitative demonstration that $Q \propto V$

(*b*) Explain why this graph indicates that $Q \propto V$ for this capacitor.
(*c*) Determine the value of the capacitance from the gradient of the graph.
4.3 The capacitor in Figure 4.19 is charged when the switch connects K to X.

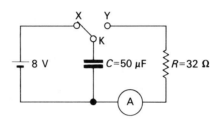

Figure 4.19

(*a*) Calculate the total charge placed on the capacitor C.
(*b*) Find the initial current during discharge when K is joined to Y.
(*c*) What is observed on the ammeter A?
(*d*) What difference would be observed on the ammeter if the value of R was increased?
4.4 The resistance of resistor R is carefully adjusted during the experiment to keep a constant meter reading of 0.2 mA. The 6 V supply is connected and the p.d. across the capacitor, measured on the voltmeter V, is observed to increase from 1.6 V after 10 s to 5 V after 30 s (Figure 4.20).

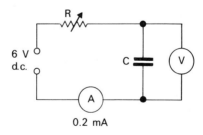

Figure 4.20

(*a*) Calculate the charge placed on the capacitor C after 10 s and 30 s, and estimate the capacitance of the capacitor.
(*b*) Suggest a suitable range for the resistor R.

B: Inductance

4:B.1 *Electromagnetic Induction*

An electric current may be produced in a closed circuit by *changes* in its magnetic environment. That is, an e.m.f. may be *generated* across a conducting wire when the magnetic field *changes*. Since it is easy to show the presence of a current in a closed circuit by the inclusion of an ammeter, the following discussion will be in terms of the induced current. However, it should be remembered that an e.m.f. is generated in order to cause a current to flow. Even if the circuit is not complete an e.m.f. would still be generated.

(1) When a conductor moves relative to a magnetic field (either by moving the conductor or moving the magnetic field so that the conductor cuts across the lines of magnetic flux), a current is produced in the conductor.

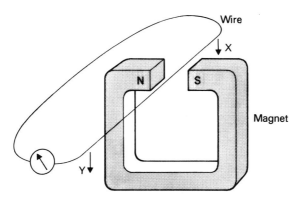

Figure 4.21

For example, if the wire is moved downwards between the poles of the magnet, as indicated in Figure 4.21, then a galvanometer connected to the ends of the wire would show a current. The conventional current direction would be from X to Y (and electron flow is from Y to X).

Notice that the field, the direction of motion, and the direction of the current are all at right angles to each other.

Fleming's right-hand (dynamo) rule* may be used to determine the direction of the current induced.

The forefinger, second finger, and thumb are held mutually perpendicular to each other with:

> Forefinger in the direction of the field.
> Thumb in the direction of the **motion.**
> Second finger gives the direction of **c**onventional current.

It may be observed that the direction of the induced current in the wire is such that the interaction of the magnetic field around the wire and the magnetic field

---

*For electron flow use the *left* hand, and the second finger will give the induced electron flow direction.

of the magnet is repulsive so that work must be done to move the wire downwards. It is this work which is converted into electrical energy. (See also (2) and Lenz's Law below.)

Note: The current flows *only* while the motion is taking place; a conductor lying across a constant magnetic field but not moving will *not* produce a current.

The *magnitude* of the induced e.m.f., and hence the value of the induced current, will increase if

the strength of the magnitude field increases,
the relative movement increases, or
the length of the conductor in the field increases.

(2) Current will also be produced in a conducting circuit if the magnetic field through the circuit changes.

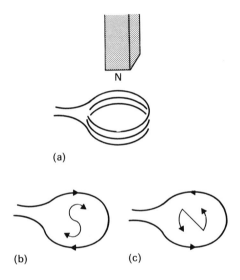

Figure 4.22   Electromagnetic induction

For example, Figure 4.22(a) shows the N pole of a magnet being pushed down into a coil of wire. The field through the conducting circuit is changing while the magnet is moving and current flows in the coil during this period. Figures 4.22(b) and (c) show the possible directions of the conventional current. In Figure 4.22(b), the current is such that the top of the coil would be a S pole, whose attraction for the magnet's N pole would help the movement and no further pushing would be required! In Figure 4.22(c), the induced current makes the top of the coil a N pole which would tend to repel the magnet's N pole; this repulsion would require work to be done to overcome it and it is this work which is converted into electrical energy. Only the conditions in Figure 4.22(c) will satisfy the Law of Conservation of Energy, hence this direction of flow must be the *correct* one. This illustrates Lenz's Law (see below).

The *magnitude* of the induced e.m.f., and hence the induced current, will increase if

> the speed of the magnet increases,
> the number of turns of the coil increases, or
> the strength of the magnet increases.

In both these examples the *direction* of the induced e.m.f., and hence the direction of the induced current, has an associated magnetic field direction which *opposes* the change.

Lenz's Law
  The direction of an induced current is always such that it opposes the change causing it.

Note: This is a result of the Law of Conservation of Energy.

The dynamo — a.c. generator

In a dynamo a coil is continuously rotated in a magnetic field producing an alternating current because the direction of the induced e.m.f. will be reversed every half revolution. This principle has great importance in the generation of electricity.

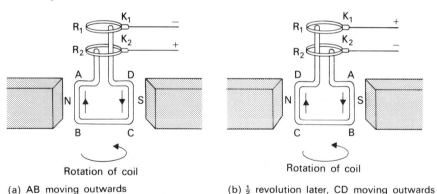

(a) AB moving outwards   (b) $\frac{1}{2}$ revolution later, CD moving outwards

Figure 4.23 a.c. generator

Note: The direction of the current induced in the coil ABCD is reversed every half revolution generating an alternating p.d. between K_1 and K_2. In many commercial generators the magnet is rotated, the coil remaining stationary.

Back e.m.f. of a motor

A motor has a coil suspended inside a magnetic field. When the coil is rotating it will 'cut' the lines of magnetic flux. An e.m.f. will then be induced across the coil *opposing* the current already flowing. This e.m.f. is termed a back e.m.f. Thus if a p.d. V is applied to a coil of resistance R the *initial* current will be V/R. But when the coil is rotating the back e.m.f. will oppose the applied p.d. and the actual *working* current will be much smaller:

$$V - \text{back e.m.f.} = I_{\text{working}} \times R$$

4:B.2 *Transformer*

If a d.c. source is connected at PQ when S is closed, a deflection is obtained on the ammeter which quickly falls to zero. An induced current is only obtained in the right-hand coil (secondary) when the current in the left-hand coil (primary) is *changing*.

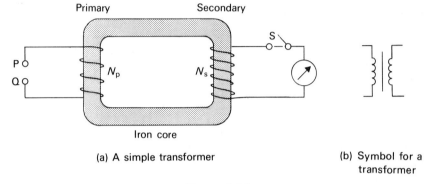

Primary   Secondary

N_p N_s

Iron core

(a) A simple transformer

(b) Symbol for a transformer

Figure 4.24

If the d.c. source is replaced by an a.c. source at PQ, a continuously changing current flows in the primary and an a.c. ammeter shows a continuous a.c. in the secondary. The primary current has set up a changing magnetic field which produces an a.c. in the secondary, opposing the field which sets it up.

The main use of a transformer is to step-up or step-down an applied p.d. or e.m.f.

For an ideal transformer, with no magnetic leakage (that is all the lines of magnetic flux threading the primary coil also thread the secondary), it may be shown that:

$$\frac{\text{Secondary p.d.}}{\text{Primary p.d.}} = \frac{\text{No. of secondary turns}}{\text{No. of primary turns}}$$

$$\frac{V_s}{V_p} = \frac{N_s}{N_p}$$

When there is *no* power loss:

$$\text{Power in primary circuit} = \text{power in secondary circuit}$$
$$V_p I_p = V_s I_s$$

However, few machines are perfectly efficient. To determine the secondary current when there *is* power loss, the efficiency formula is used.

$$\text{efficiency} = \frac{V_s I_s \text{ (power out)}}{V_p I_p \text{ (power in)}} \times 100\%$$

The power losses of the transformer are due to:

(1) $I^2 R$ loss (Joule heating).
The current flowing through the wire tends to heat up the windings. This can be decreased by reducing the current and choosing wire of suitably low resistance.

(2) Eddy currents cause power loss.
Because the iron core is itself a conducting material in a changing magnetic field, induced currents will be produced. Since they have a complete return path, they too result in heat loss. This effect may be reduced by using a laminated core.

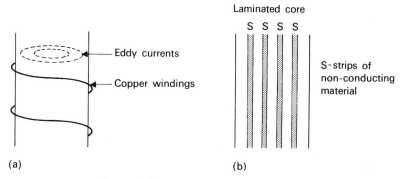

Figure 4.25 Reduction of eddy currents

(3) Hysteresis loss.
In one cycle the core has to be magnetized, demagnetized, and remagnetized in the opposite direction. Energy is required to do this. This loss may be partly overcome by using special alloys for the core.

Note: The power required by the secondary $(I_s V_s)$ will determine the power supplied by the primary $(I_p V_p)$. The values of V_p and V_s are determined by the supply and the turns ratio. Thus the value of the secondary current will affect the value of the primary current. If I_s increases, I_p must also increase.

If the secondary circuit is *open* and an a.c. supply connected to the primary, then the primary windings are effectively an inductor with reactance X_L (see Section 4:C.1), and the primary current will be *small*. When the secondary circuit is complete and supplying a current to a load, then I_p will be *larger*. However, if an equivalent d.c. supply is substituted, the only opposition to the current would be the comparatively small resistance of the windings, and a very *much larger* current would flow in the primary. Remember that *no* current would flow in the secondary with a d.c. supply.

Experiment 4.7

A number of experimental kits are available for demonstrating electromagnetic induction, Lenz's Law, the transformer, and its eddy currents.

4:B.3   *Inductors and Inductance*

An **inductor** is made of a coil of wire wound on a cylindrical frame. Inside the cylinder a **core** of magnetic material, e.g. soft iron, may be placed.

Symbol for an inductor:

With core   Without core

Induced e.m.f. produced by inductors

When the current through a coil varies an e.m.f. is induced in that coil.

Referring to Figure 4.26(a), when the switch is closed the current slowly increases from zero to a maximum value. Figure 4.26(b) shows the magnetic field B as the current flows through the coil. The magnetic field B will be increasing as the current builds up, and therefore an e.m.f. will be induced opposing I and tending to *prevent* its build up. As this e.m.f. is in the opposite direction to the

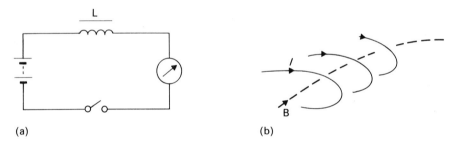

(a)   (b)

Figure 4.26   Back e.m.f.

applied voltage and current it is called the **back e.m.f.** (Also, if the supply is short-circuited, the current will fall to zero and the induced e.m.f. will *oppose* this fall, tending to maintain the current I. This time the induced e.m.f. acts in the same direction as I.)

Note: A *changing* current produces an induced e.m.f. when there is an inductor in the circuit.

The inductor will produce a *larger* induced e.m.f. if the current through it is changing *more rapidly*.

Inductance

Unit: henry (H), scaler.
The self inductance L of a coil is a measure of the amount of induced e.m.f. produced by a changing current.

induced e.m.f. = inductance × rate of change of current

V	H	$A\,s^{-1}$

Definition of the henry

If a rate of change of current of one ampere per second is accompanied by an induced e.m.f. of one volt the inductance is one henry.

Factors affecting the inductance of a coil

The induced e.m.f. is increased if the *size* of the magnetic field is increased. The design of the inductor determines the size of this field for a given current. This determines its inductance.

Experiment may show that a coil with a large number of turns will produce a larger magnetic field. A core of strong magnetic material (high relative permeability μ_r)* will also increase the magnetic field.

The magnitude of the inductance L of a coil will *increase* if
the number of turns are *increased* or
a magnetic core is present, (μ_r is increased).

Energy stored in an inductor

When a current flows in an inductor some *energy* is *stored* in the magnetic field. As the current builds up, so the value of the magnetic field increases. This energy stored in an inductor associated with the magnetic field can be a source of e.m.f. (see Section 4:B.5).

4:B.4 *Build-up and Decay of Current in Inductors*

Experiment 4.8 *To show the build-up of current in inductors*

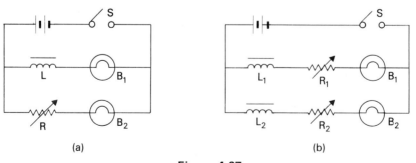

(a) (b)

Figure 4.27

* μ_r, the relative permeability, is not specifically asked for in the Higher Physics syllabus.

Using the circuit shown in Figure 4.27(a), S is closed and R adjusted until both bulbs are equally bright. S is then opened. The switch is then closed and the bulbs observed carefully. It is noticed that B_2 lights up immediately, but B_1 gradually lights up because of the back e.m.f. in the inductor.

Using the circuit shown in Figure 4.27(b), S is again closed and R_1 and R_2 adjusted until the bulbs are equally bright, then S reopened. The bulbs are then observed when S is closed. This is repeated with other inductors of different numbers of turns and cores, that is different inductances L. It is observed that the bulb with the larger inductor takes longer to light up.

The variable resistors are needed in these circuits because every inductor has its own associated resistance.

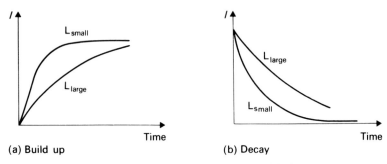

(a) Build up   (b) Decay

Figure 4.28   Variation of current with time

A graph showing the growth of current in a d.c. circuit containing an inductor is given in Figure 4.28(a). A large inductance produces a large back e.m.f. so in overcoming this, the current will take longer to build up. Also a large inductance will prevent the current falling quickly to zero, because the back e.m.f. is now in the same direction as the applied current, and it will tend to maintain the current.

If the supply shown in Figure 4.27(a) was short-circuited then removed, the bulb B_2 would go out immediately, but the bulb B_1 would dim gradually. The inductance will now prevent the current falling because the induced e.m.f. always opposes the *change*. Hence now it acts in the *same* direction as the applied current and will tend to maintain the current. Figure 4.28(b) shows the decay of the current with different inductors.

Note: Currents take *time* to build up or decay with inductors in the circuit. The larger the inductor the greater the time required.

4:B.5   *Uses of Inductors*

An inductor is sometimes referred to as a **choke**. An inductor can smooth or 'choke' variations in direct current.

Use of an inductor in smoothing

Consider an inductor connected in *series* with the 1 kΩ resistor of Figure 4.16(e), page 112. When the current decreases, the induced e.m.f. will tend to prevent the decay. Thus the remaining ripple will be smoothed out.

Use of an inductor to produce a high e.m.f.

If the current changes very rapidly in an inductor, a large e.m.f. can be produced.

Experiment 4.9

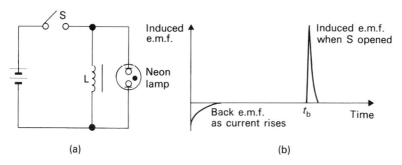

(a)   (b)

Figure 4.29

The switch S of Figure 4.29(a) is closed. There is a back e.m.f. as the current rises. Even after some time, when the current has reached its maximum, the p.d. is *not* sufficient to light the lamp. At time t_b the switch is opened. The current falls to zero immediately, since the circuit is no longer complete. A very large e.m.f. is induced for a short time in the same direction as the supply. Remember, the induced e.m.f. opposes the *change* in current. This large induced e.m.f. causes the lamp to flash. Also, in practice there may be sparking at the switch terminals. The energy to produce this e.m.f. comes from the energy stored in the magnetic field.

The other use of an inductor as a means to **block** a.c. signals, but transmit d.c. or low frequency signals is given on page 128.

Problems
4.5

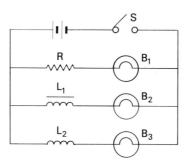

Figure 4.30

S has been closed for a few minutes and it is observed that Bulbs B_1 and B_2 are equally bright but B_3 a little dimmer. S is then opened. The bulbs are observed immediately after S is reclosed. B_1 lit up first, followed by B_3, then finally B_2. Comment on the inductors L_1 and L_2 in terms of their inductance values, number of turns, cores, and resistances, giving reasons.

4.6 A badly designed step-down transformer with a turns ratio of 6:1 has a high power-loss due to eddy currents.

(a) If the primary p.d. is 240 V what is the value of the p.d. across the secondary circuit?

(b) The current in the primary is 2 A. What may be deduced about the value of the secondary current?

(c) How may these eddy currents be reduced and what effect would this have on the answers to (a) and (b)?

C: Impedance

4:C.1 Effects of the Three Circuit Components—R, C, and L—in d.c. and a.c. Circuits

Resistor	─wwww─	Resistance	ohms
Capacitor	─┤├─	Capacitance	farads
Inductor	─ᴍᴍ─	Inductance	henries

Experiment 4.10 *Effects with d.c. supply*

The resistor, capacitor, and inductor are each connected to a d.c. source, as shown in Figure 4.31.

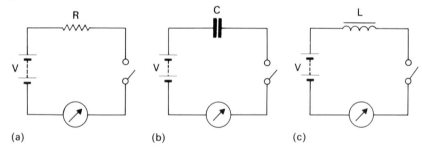

Figure 4.31

The steady current I is recorded as the e.m.f. of the source is altered in each circuit.

In (a) the ratio V/I is almost constant and equal to the resistance R of the resistor providing that the internal resistance of the battery is negligible compared with R.

In (b) the current I is zero, after the initial charging current has died away, for all values V of the d.c. source. Hence V/I is infinite showing that a capacitor is a break in a d.c. circuit.

In (c) the current I is very large, after the initial build up, and therefore the ratio V/I is very small. An inductor usually has a small associated resistance, but otherwise it does not oppose a direct current.

The only opposition to a direct current comes from a *resistor*. A capacitor acts as a break in a d.c. circuit.

Experiment 4.11 *Effects with a.c. supply*

The d.c. sources in Figure 4.31 are replaced with a.c. sources (Figure 4.32).

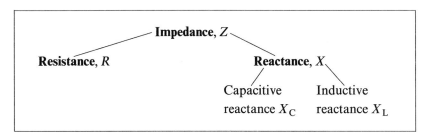

Figure 4.32

The current I is recorded on the a.c. ammeter for different values of the alternating supply V in each circuit.

In (a) the ratio V/I is again constant and equal to the resistance R.

In (b) finite current readings are obtained as charge is flowing continuously on and off the capacitor every cycle of the applied alternating e.m.f. There is, however, some opposition to the current as any charge on the plates of the capacitor tends to repel subsequent charge. The ratio V/I is found to be a constant for the given capacitor and is called its **reactance** X_C.

In (c) lower current readings than with equal values of direct e.m.f. are obtained, showing that an inductor opposes the current flow in a.c. circuits. A back e.m.f. is induced, as the current flow is continuously changing, which opposes the applied alternating e.m.f. The ratio V/I is a constant for the given inductor and is called its **reactance** X_L.

The general term for opposition to current flow is **impedance** Z. The impedance of a resistor is called **resistance** R, but the impedance of a capacitor or inductor is called **reactance** X.

Impedance, Z

Resistance, R   **Reactance, X**

Capacitive   Inductive
reactance X_C reactance X_L

All the quantities Z, R, X are measured in *ohms*, but resistances and reactances *cannot* be added together to give the overall impedance.

To summarize:

	d.c.	a.c.
Resistor $R = \dfrac{V}{I}$	$Z = R$	$Z = R$
Capacitor $X_C = \dfrac{V}{I}$	$Z = \infty$	$Z = X_C$
Inductor $X_L = \dfrac{V}{I}$	$Z = 0$	$Z = X_L$

Notice that for a pure inductor Z is zero in a d.c. circuit but due to the resistance of the wire, small finite values of Z will be obtained in practical experiments. For a capacitor in a d.c. circuit Z is infinite, ∞, because the current flow is zero.

4:C.2 Factors affecting Capacitive Reactance X_C

Experiment 4.12

A capacitor is connected in series with an ammeter to an a.c. supply as in Figure 4.33.

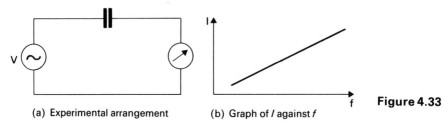

(a) Experimental arrangement   (b) Graph of I against f **Figure 4.33**

The e.m.f. of the source is maintained at a constant value V and the current recorded for:

(a) different values of capacitance C and
(b) different frequencies f of the a.c. supply.

The reactance X_C is equal to V/I. For a constant value of V an increase in current I implies a decrease of reactance X_C.

Note: A larger current means a smaller opposition to that current.

In (a) the current is observed to increase as the capacitance C is increased, hence a larger capacitance has a smaller *opposition* to the current, therefore a smaller reactance X_C. This larger capacitance is able to accommodate more charge and the current can flow more easily on and off the plates.

In (b) the current I also increases as the frequency is increased. A graph of current against frequency, shown in Figure 4.33(b) indicates that $I \propto f$ for the capacitor. At high frequencies the capacitor does not have time to become fully charged, hence larger charging currents are still flowing when the direction of the applied e.m.f. reverses. At low frequencies the capacitor may be fully charged with negligible currents flowing when the direction of the e.m.f. reverses. Thus at high frequencies the average current is large giving a small reactance X_C.

$$I \uparrow \text{ as } C \uparrow \qquad X_C \downarrow \text{ as } C \uparrow$$
$$I \uparrow \text{ as } f \uparrow \qquad X_C \downarrow \text{ as } f \uparrow$$
$$\text{It may be shown that } X_C = \frac{1}{2\pi f C}$$
$$\uparrow \text{ increases } \quad \downarrow \text{ decreases}$$

Use of a capacitor to block d.c. but pass a.c.

When the frequency increases, the value of X_C becomes quite small. This property makes the capacitor very useful in a.c. circuits to block d.c. signals.

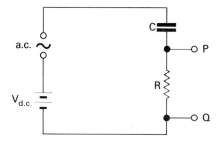

Figure 4.34

Referring to the circuit of Figure 4.34 the capacitor will charge up until the p.d. across it is $V_{d.c.}$. The d.c. supply will then cease to provide current in the circuit. However, an alternating current will still flow on and off the capacitor plates and through the resistor R. Thus there is an alternating p.d. across R. An output at PQ will be *alternating*. The capacitor has effectively **blocked** the d.c. and passed on the a.c. signal. In this arrangement it is called a *coupling* capacitor since it couples or 'joins' the a.c. signal to the output.

To increase the value of the alternating potential at PQ the impedance of C should be low compared to the resistance of R, $(V_R = IR$ and $V_C = IX_C)$.

The coupling capacitors C_1 and C_2 of Figure 3.9, page 87 have this function of passing on the a.c. signals but blocking the d.c. components. However, the capacitor C_3 of this figure provides a low impedance path to *unwanted* high frequency pulses and is called a *decoupling* capacitor.

4:C.3 *Factors affecting Inductive Reactance X_L*

Experiment 4.13

An inductor is connected to an a.c. supply, as in Figure 4.35.

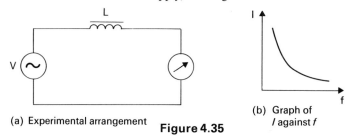

(a) Experimental arrangement   **Figure 4.35**   (b) Graph of I against f

The e.m.f. is maintained at a constant value V and the current recorded for:

(a) different inductances L and
(b) different frequencies f of the a.c. supply.

As in Experiment 4.12 a smaller current I implies a larger opposition to that current, hence a larger value of X_L.

In (a) the current decreases as the inductance increases. A larger inductance produces a larger back e.m.f. hence more opposition to the current, that is a larger reactance X_L.

In (b) the current I also decreases as the frequency increases. The graph of current against frequency shown in Figure 4.35(b) *suggests* an inverse relationship between current and frequency. If a graph of I and $1/f$ is plotted, a straight line through the origin is obtained indicating that $I \propto 1/f$. A more rapid *change* in applied current will increase the size of the back e.m.f. opposing the current, giving a larger reactance X_L.

$$I \downarrow \text{ as } L \uparrow \qquad X_L \uparrow \text{ as } L \uparrow$$
$$I \downarrow \text{ as } f \uparrow \qquad X_L \uparrow \text{ as } f \uparrow$$

It may be shown that $X_L = 2\pi f L$

↑ increases ↓ decreases

Notice the opposite effect, on the reactance, of the frequency with an inductor, as opposed to a capacitor. The inductor discourages the passage of a.c. since X_L rises as the frequency increases.

Note: A reactance X varies with frequency. A resistance R does not change with frequency.

Use of an inductor to block a.c. but transmit d.c.

The input V, shown in Figure 4.36(a), contains a steady direct voltage with high frequency alternating signals. X_L is large at high frequency, hence the a.c. flowing through L and R will be small. The inductor will *block* the a.c. signal. If the resistance of the inductor coil is much less than the resistance of R, then the p.d. across R will be almost equal to the direct voltage of the supply. Observe that the p.d. across L will be almost entirely alternating.

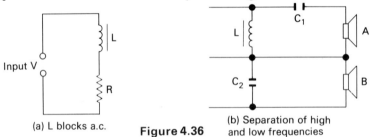

(a) L blocks a.c.   **Figure 4.36**   (b) Separation of high and low frequencies

An inductor can be used with capacitors to *separate* high and low frequency signals, see Figure 4.36(b). Any high frequency signal will be blocked by the inductor. These signals will be routed to loudspeaker A via C_1 and C_2, since X_C is small at high frequencies. The low frequency signals will take the now lower impedance path via L to loudspeaker B.

Note: A capacitor can block d.c. signals but an inductor blocks a.c. signals.

4:C.4   *Lagging and Leading*

Resistor

With a **resistor** there is no effect with an alteration of frequency of an a.c. source. The current and p.d. across a resistor reach their maximum values at the same time. They are **in phase**.

Capacitor

When the switch is closed the current falls from an initial maximum value to zero, while the p.d. across the capacitor rises from zero to a maximum, (Figure 4.37).

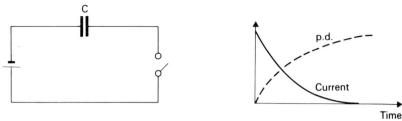

Figure 4.37 Current and p.d. phase difference

With an a.c. supply the same process is repeated every cycle, (Figure 4.38).

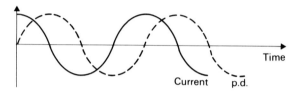

Figure 4.38 Current leads p.d.

For a pure capacitor:
The current **leads** the p.d. by 90° ($\frac{1}{4}$ cycle).*
The current and p.d. have a **phase difference** of 90°.

Inductor

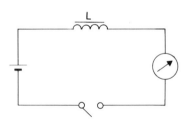

Figure 4.39

After the switch is closed the current takes time to reach its maximum value due to the back e.m.f., but the p.d. across the inductor attains its maximum value immediately, (Figure 4.39).

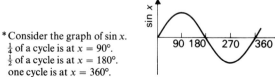

*Consider the graph of sin x.
$\frac{1}{4}$ of a cycle is at $x = 90°$.
$\frac{1}{2}$ of a cycle is at $x = 180°$.
one cycle is at $x = 360°$.
So for any oscillation $\frac{1}{4}$ of a cycle may be made equivalent to 90° by a suitable choice of axes.

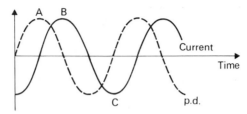

Figure 4.40 Current lags p.d.

With an a.c. supply when the p.d. is increasing in the forward ($+$ve) direction there will be a back e.m.f. preventing the current reaching its maximum value. But when the p.d. is decreasing the induced e.m.f. will now tend to maintain the current in the forward direction (maximum at B), see Figure 4.40. As the p.d. increases in the backward ($-$ve) direction the current will follow, but again due to the induced e.m.f. it will not reach its maximum until the p.d. falls to zero again (at C).

For a pure indicator:
The current **lags** the p.d. by 90° ($\frac{1}{4}$ cycle).
The current and p.d. have a **phase difference** of 90°.

Note: It is because of these phase differences between current and p.d. with capacitors and inductors that reactances and resistances *cannot* be simply added to give impedances. (See Section 4:C.1.)

4:C.5 *Circuits containing a Capacitor and an Inductor**

A circuit containing a capacitor and an inductor can produce electrical oscillations. A capacitor charged by a d.c. supply can be discharged through an inductor to give a damped a.c. signal.

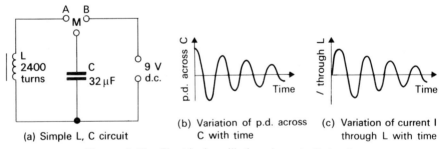

(a) Simple L, C circuit   (b) Variation of p.d. across C with time   (c) Variation of current I through L with time

Figure 4.41 Electrical oscillations in an L, C circuit

M is joined to B, see Figure 4.41(a), to charge the capacitor, and then joined to A to discharge the capacitor through the inductor.

* Section 4:C.5 is only for interest or the Special Topics.

A charged capacitor has potential energy, but when it discharges through an inductor the current is a form of kinetic energy. While the current is flowing through the coil it produces a magnetic field which initially opposes the charge flowing. Because the flow of charge, the current, *decreases* when a capacitor is discharged, the magnetic field tends to collapse but the induced e.m.f. will tend to *maintain* the field and keep the current flowing. Therefore the capacitor is recharged with the opposite polarity (Figure 4.42). The energy is now once again potential energy in the capacitor. The process is repeated but with decreasing amplitude because some of the energy is converted into heat.

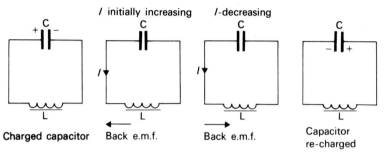

Figure 4.42 Back e.m.f. assists in recharging the capacitor

The natural frequency depends on the values of L and C. It is observed that if either L or C are *increased* the natural frequency *decreases*.

The natural frequency f_0 is such that the reactance of C equals the resistance of L.

$$\frac{1}{2\pi f_0 C} = 2\pi f_0 L$$

$$f_0^2 = \frac{1}{4\pi^2 CL}$$

To overcome damping, energy must be fed in at the resonance frequency to recharge the capacitor every cycle. There are three methods of maintaining the oscillations.

(1) *Parallel tuned circuits*; where the L and C components are in parallel with the a.c. supply, (Figure 4.43(a)).
(2) *Series tuned circuits*; where the L and C components are in series with the a.c. supply, (Figure 4.43(b)).
(3) *Oscillators*; where an amplifier is used to feed back part of the oscillations.

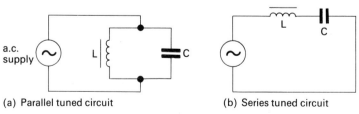

(a) Parallel tuned circuit   (b) Series tuned circuit

Figure 4.43 Circuits to maintain electrical oscillations

4:C.6 *Energy dissipation in R, C, and L*

> Energy is dissipated in a resistor only.

Resistor

Energy is always being taken from the source, either d.c. or a.c., and changed into another form of energy, heat for example.

Capacitor

With an alternating supply, the positive and negative half cycles are similar. On average no energy is taken from the source. The capacitor is charged and energy stored but this is returned to the source during discharging.

Pure inductor

With an alternating supply, energy is stored in the magnetic field during current build up and returns to the source of supply when the field collapses. Averaged over each cycle no energy is taken from the source.

4:C.7 *Summary of the Three Circuit Elements*

	Resistor Resistance (R) ohm (Ω)	Capacitor Capacitance (C) farad (F)	Inductor Inductance (L) henry (H)
Symbol	—\/\/\/—	—||—	—⌒⌒⌒—
Factors affecting the value	$R = \dfrac{\rho l}{A}$	$C \propto \dfrac{A\varepsilon_r}{d}$ (parallel plate)	$L\uparrow$ with number of turns $L\uparrow$ if a core is present
Impedance Z in an a.c. circuit (unit: ohm)	R	Reactance X_C	Reactance X_L
Dependence of Z on a.c. frequency	R independent of frequency	$X_C = \dfrac{1}{2\pi f C}$	$X_L = 2\pi f L$
Initial effects in d.c. circuit	I attains full value immediately	C takes *time* to charge up to full p.d.	I takes *time* to build up, due to back e.m.f.
Phase	V and I in phase	I leads V by 90°	I lags V by 90°

Problems

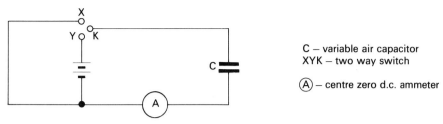

Figure 4.44

C — variable air capacitor
XYK — two way switch

(A) — centre zero d.c. ammeter

4.7 Considering the above circuit, sketch graphs on the same axes to show how the current varies with time when the following are performed:
(a) K is joined to Y.
(b) K is joined to X.
(c) The vanes of the capacitor are adjusted to their positions of maximum overlap and K is joined to Y.
(d) the capacitor is immersed in a non-conducting liquid of dielectric constant $\varepsilon_r = 40$ and K is joined to Y.
Sketch similar graphs on another set of axes when:
(e) C is replaced by an inductor L with a core and KY is closed.
(f) the core is removed from the inductor and KY is closed.
Comment briefly on the graphs (a) to (f).

4.8 In the circuit in Figure 4.45 V is a steady a.c. source. What is the effect on the current when the frequency is increased from 50 to 500 hertz with S_1 closed. What is the difference, if any, if both S_1 and S_2 are closed?

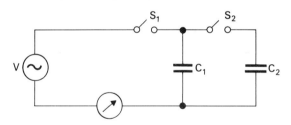

Figure 4.45

4.9 A student conducts an experiment with a supply of constant e.m.f. (Figure 4.46(a)), but gradually increases the frequency.

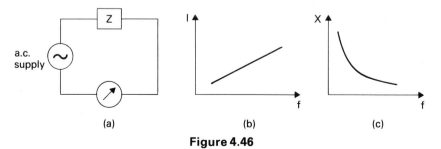

(a)   (b)   (c)

Figure 4.46

Is Z a resistor, capacitor or an inductor if:

(*a*) the reading on the ammeter is observed to decrease,

(*b*) the reading on the ammeter remains constant,

(*c*) the reading on the ammeter is observed to increase.

(*d*) For one of (*a*) or (*c*) he plots graphs of *I* against *f* and reactance *X* against *f* and obtains the graphs shown in Figure 4.46(b) and (c). Explain whether the graphs refer to (*a*) or (*c*).

4.10 Assume that the c.r.o.s. in Figure 4.47 have been calibrated to give the same response to the same potential difference. Briefly explain the trace observed on each c.r.o. when the a.c. input is (*a*) a high frequency, (*b*) a low frequency.

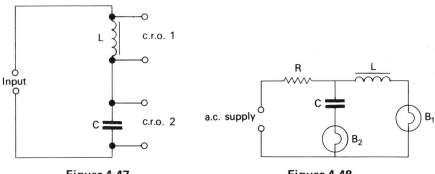

Figure 4.47   **Figure 4.48**

4.11 Explain briefly which lamp in Figure 4.48 would be brighter at (*a*) low frequencies, (*b*) high frequencies.

4.12

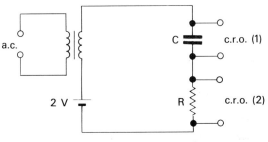

Figure 4.49

The a.c. supply of peak value 10 V is connected to a 2:1 step-down transformer (Figure 4.49). C has a large value of capacitance and R is a 10 kΩ resistor. Sketch with numerical values the traces observed on c.r.o. (1) and c.r.o. (2). Briefly explain your answer.

5 Optics and Waves

A: Reflection and Refraction

5:A.1 Reflection

When a narrow, parallel beam of light, a **ray**, is shone on to a mirror the light is reflected such that the angle of incidence is equal to the angle of reflection, as shown in Figure 5.1.

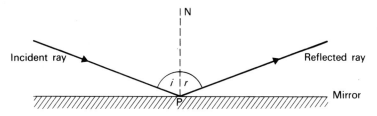

Figure 5.1 Reflection

NP —normal at the point of incidence P (a normal is a line drawn at right angles to a second line or surface).

i —angle of incidence between the incident ray and the normal.

r —angle of reflection between the reflected ray and the normal.

Laws of reflection:

(1) Angle of incidence i = angle of reflection r
(2) The incident ray, reflected ray, and the normal at the point of incidence are all in the same plane.

Note: If the light were replaced by a stream of ball bearings travelling in the direction of the incident ray and a hard board were substituted for the mirror, the balls would bounce off in the direction of the reflected rays. This suggests that a ray of light may also be regarded as a stream of 'particles', called **photons**, travelling in straight lines.

Experiment 5.1 *Reflection with plane mirrors*

Using a ray box and a single incident ray directed onto a plane mirror, the first law of reflection can be verified for a range of angles of incidence. Remember to measure angles between the rays and the *normal*.

A **ray diagram** can be drawn to determine the position of the image of an object formed in a plane mirror.

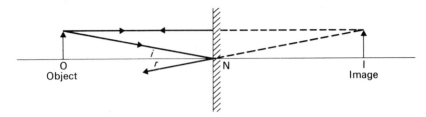

Figure 5.2 Image position in a plane mirror

Consider the two rays shown:
(1) A ray perpendicular to the mirror is reflected back along its incident path.
(2) A ray drawn to any point N with angle of incidence i will be reflected such that $i = r$.
The tip of the object will be at the point where these two rays appear to come from. By extending the rays behind the mirror they meet at the tip of the image I. The geometry shows ON = NI.

The nature of the image of this object is: upright, virtual (it cannot be obtained on a screen), and the same size. It is also *laterally* inverted: SUN will appear ꞂUꙄ.

5:A.2 *Refraction*

Refraction, the change in direction of light at a boundary between two media, is due to the difference in the velocity of the light in the two media.

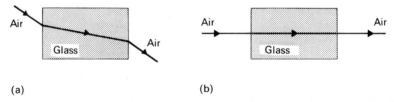

(a)   (b)

Figure 5.3 Refraction

The light, that is the photons, travel *slower* in the glass than in the air. However, if the ray is perpendicular to the interface of the two media, there is no change in direction, although the light is still travelling slower inside the glass (Figure 5.3(b)).

Experiment 5.2 *To investigate the relationship between the angle of refraction and the angle of incidence*

A semicircular block of glass or perspex is used to obtain refraction at one interface only, as shown in Figure 5.4(a).

NN′—normal at the point of incidence.
θ_1 —angle of incidence (angle between the incident ray and normal).
θ_2 —angle of refraction (angle between the refracted ray and normal).

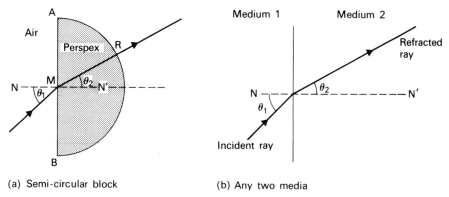

(a) Semi-circular block   (b) Any two media

Figure 5.4

A ray of light is directed to the midpoint M of the straight side AB. After refraction it traverses a radius of the block, hence there is no refraction at R.

Angles of refraction θ_2 are measured for different angles of incidence θ_1 and the following table is completed by experiment.

θ_1	θ_2	$\sin \theta_1$	$\sin \theta_2$	$\sin \theta_1/\sin \theta_2$

The experiment shows that the last column, $\sin \theta_1/\sin \theta_2$ is a constant within the limits of the experiment.

Snell's Law

For light (or any wave motion) of a given frequency and any two media,

$$\frac{\sin \theta_1}{\sin \theta_2} = \text{a constant}$$

where θ_1 and θ_2 are the angles between the ray and the normal in each medium, as shown in Figure 5.4(b).

It may be demonstrated that this constant is the **relative refractive index** $_1n_2$ for the two media.

Thus, for the above experiment, the relative refractive index for air to perspex is given by:

$$_1n_2 = \,_{air}n_{perspex} = \frac{\sin \theta_1 \text{ (angle of incidence in air)}}{\sin \theta_2 \text{ (angle of refraction in perspex)}}$$

Note: The incident ray, the refracted ray, and the normal at the point of incidence are all in the same plane.

Absolute refractive index, n

A ratio, no units.

The absolute refractive index n of a medium (usually referred to simply as the refractive index) is defined as the ratio of the speeds of light in vacuum and in the medium. A larger refractive index indicates a slower speed of light.

$$n = \frac{\text{speed of light in vacuum } v_0}{\text{speed of light in medium } v}$$

$$n = \frac{v_0}{v}$$

Note: The refractive index varies with the frequency of the light.

Relationship between speed of light in two media and the angles of incidence and refraction*

Consider a beam of light refracted from a vacuum into a medium of refractive index n and speed of light v.

In Figure 5.5, AC is perpendicular to XC and BD is perpendicular to CX′.

hence $$\hat{ACB} = \theta_0 \quad \text{and} \quad \hat{CBD} = \theta$$

When ray Y travels the distance AB in vacuum, ray X travels the distance CD in the medium.

Since the times taken to travel AB and CD are the same

$$\frac{AB}{v_0} = \frac{CD}{v}$$

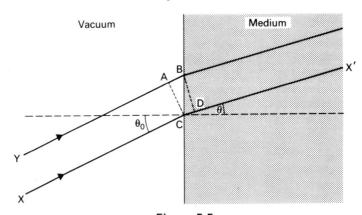

Figure 5.5

giving $$\frac{v_0}{v} = \frac{AB}{CD} = \frac{AB/BC}{CD/BC} = \frac{\sin \theta_0}{\sin \theta}$$

hence $$\frac{v_0}{v} = \frac{\sin \theta_0}{\sin \theta}$$

* This proof is not required for examination purposes.

But the refractive index is defined as $n = \dfrac{v_0}{v}$, hence for refraction from vacuum into a medium:

$$n = \frac{\sin \theta_0}{\sin \theta}$$

where θ_0 is the angle of incidence in vacuum and θ is the angle of refraction in the medium.

The (absolute) refractive index is thus a measure of the amount by which the rays of light change *direction* when they enter a medium from a vacuum. The larger the refractive index the greater this change in direction.

Examples of absolute refractive indices

Vacuum	$n = 1$	Air	$n = 1.0003$
Water	$n = 1.33$	Glass	$n = 1.50$ to 1.65
Diamond	$n = 2.41$	Perspex	$n = 1.5$

It should be noted that if the relative refractive index from air into a medium (which is comparatively easy to measure) is used instead of the absolute refractive index from vacuum to the medium, the error is very small because the refractive index of air is so close to unity.

Refraction between any two media

Provided that refraction takes place between air(vacuum) and the medium, then the above equation can be used. If refraction takes place between two different media, water and glass, for example, a relative refractive index would be required, but these are *not* listed in data books!

The above relationship between angles and speeds could be applied to refraction at a boundary between any two media 1 and 2 giving

$$\frac{v_1}{v_2} = \frac{\sin \theta_1}{\sin \theta_2}$$

also $\qquad n_1 = \dfrac{v_0}{v_1} \qquad$ and $\qquad n_2 = \dfrac{v_0}{v_2}$

hence $\qquad \dfrac{v_1}{v_2} = \dfrac{n_2}{n_1} = \dfrac{\sin \theta_1}{\sin \theta_2}$

The equation relating refractive indices and angles may be written in a symmetrical form and should be learnt.

$$n_1 \sin \theta_1 = n_2 \sin \theta_2$$

It may now be observed that the **relative refractive index** $_1n_2 = \dfrac{n_2}{n_1}$ as indicated at the end of Experiment 5.2.

From the practical point of view, refractive index is determined by measuring angles, not by measuring speeds of light!

Example: Light is refracted at a perspex/water boundary. Calculate the angle of incidence in perspex for a ray of light giving an angle of refraction of 40° in the water.

Using
$$n_1 \sin \theta_1 = n_2 \sin \theta_2$$
$$1.5 \sin \theta_1 = 1.33 \sin 40°$$

Giving the angle of incidence $\theta_1 = 34.7°$.

The formula $n = \dfrac{\sin \theta_1}{\sin \theta_2}$ must be used with care. It only applies if refraction takes place between air (vacuum) and the medium, but $n_1 \sin \theta_1 = n_2 \sin \theta_2$ can always be used. In many examples one medium is *air* and either n_1 or n_2 will be equal to 1.

5:A.3 *Critical Angle*

If light travels from a medium into air the angle of *refraction* in the air is greater than the angle of *incidence* in the medium, since the refractive index of a medium is larger than that for air, see Figure 5.6(a).

If the angle of incidence is increased until the angle in air is 90°, as in Figure 5.6(b), the light can just emerge along the interface, and the angle in the medium is called the **critical angle** C. The critical angle is defined as the angle in the *medium* when the angle in air is 90°.

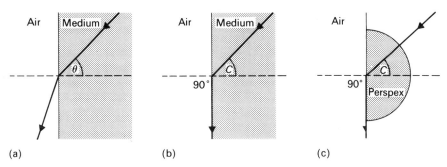

(a)   (b)   (c)

Figure 5.6   Critical angle C

From Figure 5.6(b),

$$n = \frac{\sin \theta_0 \text{ (angle of refraction in air)}}{\sin \theta \text{ (angle of incidence in medium)}}$$

or using
$$n_1 \sin \theta_1 = n_2 \sin \theta_2$$
$$n \sin C = 1 \times \sin 90°$$
$$n = \frac{\sin 90°}{\sin C}$$

$$n = \frac{1}{\sin C} \quad \text{and} \quad \sin C = \frac{1}{n}$$

Note: This equation applies only to refraction between *air* and a medium.

Examples of critical angles:

Using $\sin C = \frac{1}{n}$ the following critical angles should be calculated.

Diamond	$n = 2.41$	$C = 24.5°$
Perspex	$n = 1.5$	$C = 41.8°$
Water	$n = 1.33$	$C = 48.8°$

Experiment 5.3 *Measurement of the critical angle of perspex*

Using a semicircular block light is directed through the block at the midpoint of the straight side. The block is slowly rotated until the refracted ray lies along the perspex/air boundary, as shown in Figure 5.6(c), and the critical angle measured.

Instruments have been designed which measure this critical angle accurately, thus providing reliable values for refractive indices. An instrument which measures refractive indices is called a **refractometer**.

Total internal reflection

When the angle of incidence in the medium is greater than the critical angle, light cannot be refracted out into the air. All the light is then reflected back into the medium at the interface, as shown in Figure 5.7. This is termed **total internal reflection**.

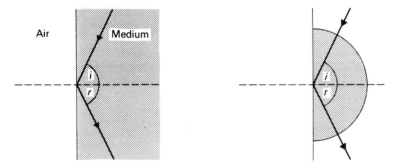

Figure 5.7 Total internal reflection

By the law of reflection, $i = r$. (Remember that reflection also takes place to some extent when the angle in the medium is *less* than the critical angle.)

Examples of total internal reflection

A 45° isosceles glass or perspex prism may be used as a mirror (Figure 5.8(a)) or

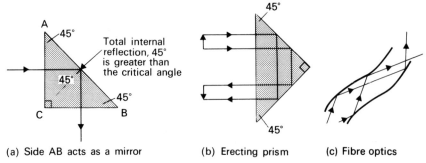

(a) Side AB acts as a mirror   (b) Erecting prism   (c) Fibre optics

Figure 5.8   Total internal reflection

as an erecting prism (Figure 5.8(b)). Erecting prisms are used in pairs in some binoculars. Unlike reflection with mirrors, there is no energy loss with total internal reflection.

In **fibre optics** light is shone into a transparent plastic fibre at a suitable angle so that total internal reflection occurs at the sides of the fibre throughout its length and the light cannot escape, (Figure 5.8(c)).

Diamonds and other precious stones have high refractive indices and therefore low critical angles. The light is often internally reflected and only emitted in certain directions, giving intense emitted beams and causing the stone to sparkle.

5:A.4   *Deviation and Dispersion*

Deviation

If light is passed through a transparent rectangular block (Figure 5.9(a)) refraction occurs when the light enters and leaves the block but the emergent ray EE' is *parallel* to the direction of the incident ray II'.

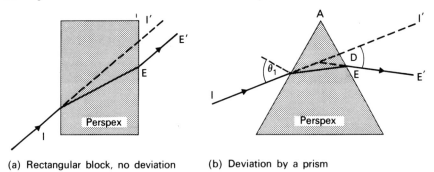

(a) Rectangular block, no deviation   (b) Deviation by a prism

Figure 5.9   Deviation

When a prism is employed (Figure 5.9(b)) the refraction causes an overall change in direction which is termed **deviation**.

The **angle of deviation D** is the angle between the incident ray direction II' and the emergent ray direction EE'.

The angle of deviation depends on: the angle of incidence θ_1, the apex angle of the prism A, and the refractive index of the prism material.

Note: Because the refractive index varies with the frequency of the light used, the angle of deviation will also vary with the frequency.

Minimum angle of deviation

Experiment 5.4

A ray of light is directed along a line II' drawn on a sheet of paper. The prism is placed in the path of the light with a small angle of incidence θ_1, as shown in Figure 5.10.

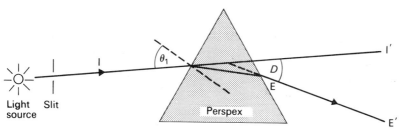

Figure 5.10 Variation of angle *D*

The prism is slowly rotated, increasing the angle of incidence θ_1. It is observed that the emergent ray EE' approaches the line II', *reducing* the angle of deviation *D*, and then moves away from II', *increasing* the angle of deviation. The angle when the deviation is *least* is called the **minimum angle of deviation**. This position of minimum deviation is important when viewing spectra.

Dispersion

It has been mentioned that the refractive index *varies* with frequency. Hence different *colours* have different refractive indices. The refractive index for blue light is greater than the refractive index for red light, hence blue is refracted more than the red, as shown in Figure 5.11.

The visible spectrum consists of those wavelengths of electromagnetic radiation which are detected by the normal eye, $\lambda = 4 \times 10^{-7}$ m to 7×10^{-7} m. If radiation of all these wavelengths overlap then white light is observed. A spectrum can be produced by a glass or perspex prism, as shown in Figure 5.11.

The deviation produced by a prism depends on the refractive index but because the refractive index varies with frequency the deviation of the blue light

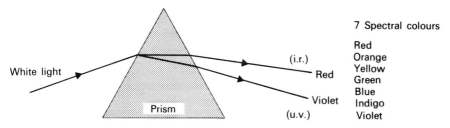

Figure 5.11 Visible spectrum

will be different to that of the red light. This difference in the deviation of the blue and red light (or of any two waves of different frequency) is called **dispersion**.

The **dispersive power** of a material is a measure of the dispersion and depends on the *difference* in the refractive indices for the blue and red light.

Note: The colour of the light depends on its wavelength. Red light has the longest wavelengths and the wavelength decreases across the spectrum to violet which has the shortest wavelengths.

A beam of light of a single colour, comprising a narrow band of wavelengths, is called **monochromatic**.

5:A.5 *Wave Optics: Refraction*

This section is concerned with the refraction of light *waves* and can be extended to other forms of electromagnetic radiation.

During refraction of waves the frequency remains *unchanged*, (see Section 5:C.1). But for waves $v = \lambda f$ hence in each medium $v_1 = \lambda_1 f$ and $v_2 = \lambda_2 f$ giving

$$\frac{v_1}{v_2} = \frac{\lambda_1}{\lambda_2} = \frac{\sin \theta_1}{\sin \theta_2}$$

for waves of speeds v_1 and v_2, wavelengths λ_1 and λ_2, and angles θ_1 and θ_2 in each medium respectively. The relationship between speeds and angles was proved on page 138. Observe that these 'ratio' equations do *not* contain the refractive index n.

Equations involving the refractive index, such as $n_1 \sin \theta_1 = n_2 \sin \theta_2$ are written in a symmetrical form.

Since $\dfrac{n_2}{n_1} = \dfrac{v_1}{v_2}$ from page 139, then

$$n_1 v_1 = n_2 v_2$$
$$n_1 \lambda_1 = n_2 \lambda_2$$

Example: A beam of infra-red radiation of wavelength $1.2\,\mu m$ is refracted by a certain glass. The angles of incidence and refraction are 30° and 21° respectively. Calculate (a) the wavelength, (b) the frequency, and (c) the speed of the radiation inside the glass. (d) Determine the refractive index for the glass for this infra-red.

(a) $\qquad \dfrac{\lambda_1}{\lambda_2} = \dfrac{\sin \theta_1}{\sin \theta_2}$ thus $\dfrac{1.2 \times 10^{-6}}{\lambda_2} = \dfrac{\sin 30°}{\sin 21°}$

Giving λ_2, the wavelength in the glass $= 0.86\,\mu m$.

(b) In air: $v_1 = \lambda_1 f$ thus $3 \times 10^8 = 1.2 \times 10^{-6} \times f$
giving the frequency in *air* $f = 2.5 \times 10^{14}\,Hz$
The frequency does not change, hence frequency in the glass $= 2.5 \times 10^{14}\,Hz$

(c) In glass: $v_2 = \lambda_2 f$ giving $v_2 = 0.86 \times 10^{-6} \times 2.5 \times 10^{14}$
$\qquad\qquad\qquad\qquad = 2.15 \times 10^8\,m\,s^{-1}$

(d) Using $\quad n_1 \sin \theta_1 = n_2 \sin \theta_2$
$$1 \times \sin 30° = n \sin 21°$$
$$n_2 = 1.4$$

The refractive index for infra-red of this glass = 1.4.

Note: Remember that the refractive index varies with frequency. This is the *cause* of dispersion. Red light has a *longer* wavelength, shorter frequency, and smaller refractive index than blue light.

Problems

5.1 The velocity of light is greatest in a vacuum. What does this imply about the possible values a refractive index may have?

5.2 (a) A ray of light passes from glass, $n = 1.5$, into water, $n = 1.33$. Draw a sketch to show the relative sizes of the angles of incidence and refraction. No calculations are required.

(b) Determine the angles of refraction in the media shown in Figure 5.12, if the incident light is in air, (where n (quartz) = 1.54 and n (glycerol) = 1.47).

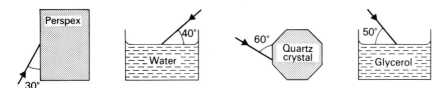

Figure 5.12

5.3 Light of frequency 6×10^{14} Hz travels from air into a block of glass of refractive index 1.5. (Assume the velocity of light in air is 3×10^8 m s^{-1}.) Calculate the velocity, wavelength, and frequency of the light in the glass.

5.4 A ray of monochromatic light passes from water into glass. The angle of incidence in the water is 50°. The speeds of light in the water and glass are 2.26×10^8 m s^{-1} and 1.7×10^8 m s^{-1} respectively.

(a) Calculate the angle of refraction in the glass.

(b) In which medium, water or glass, is the wavelength the longer?

(c) Compare the frequency in the water and glass.

(d) State the speed of light in air, and calculate the wavelength in the glass if the wavelength in the air is 580 nm.

5.5 A ray of white light is dispersed by the prism in Figure 5.13, giving a spectrum at S. The angle x is found to be 0.7°. If the refractive index of the prism material for red light is 1.51, what is its refractive index for the blue light?

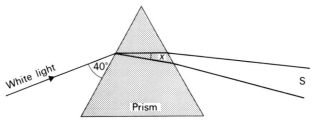

Figure 5.13

B: Lenses, Illuminance and Optical Instruments

5:B.1 *Refraction through Lenses*

Parallel rays of light are refracted by a convex lens in such a way that they converge to a **real principal focus** F. Such a lens is often called a **converging lens**.

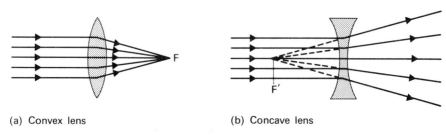

(a) Convex lens　　　　　　　　(b) Concave lens

Figure 5.14 Lenses

For a concave lens the parallel rays of light are refracted outwards by the lens in such a way that they appear to diverge from a **virtual focus F′**. This lens is termed a **diverging lens**. The distance of F or F′ from the centre of the lens is called the **focal length**.

Experiment 5.5 *Measurement of focal lengths*

For a thin **converging lens**, the focal length can be determined by obtaining a clear image of a distant object on a blank wall. The distant object could be a building outside the window on the opposite side of the room. Since rays from a *distant* object are effectively parallel, the image is formed at the focus of the lens. Thus the distance of the lens from the wall is the focal length. Also it is observed that this image is inverted and diminished. This method is very convenient but not of great accuracy. Because diverging lenses have virtual foci this method cannot be applied to them.

Location and nature of images using ray diagrams

The **principal axis** is a line through the midpoint M of the lens (Figure 5.15(a)). By convention the rays are drawn from left to right. O is the position of the object and I the image position.

> The following *two* rays are sufficient for locating the image position.
> (1) A ray parallel to the principal axis passes through the focus after refraction.
> (2) A ray passing through the centre of the lens continues straight on undeviated.

In Figure 5.15(a) X is a point on the object. The two rays (1) and (2) above are constructed. They meet at point Y on the image. Y is the point on the image which corresponds to X on the object.

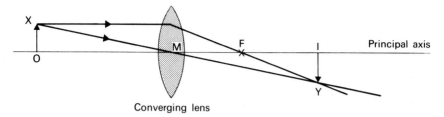

Converging lens

(a) The two rays required for image location

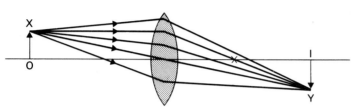

(b) All rays from X meet at Y

Figure 5.15 Converging lens

The distance FM is called the **focal length** f.
The distance OM is called the **object distance** u.
The distance IM is called the **image distance** v.

Note: Any other ray passing through point X will also be refracted by the lens to Y, Figure 5.15(b).

The image is a **real image** if it may be shown on a screen placed at I, as in Figure 5.15. It will be on the *opposite* side of a single lens from the object.

The image is a **virtual image** if it cannot be obtained on a screen, as in Figure 5.16. It is the position from which the rays of light *appear* to diverge. It is not essential for light to pass through the virtual image position in order to form such an image. The virtual image is on the *same* side of a converging lens as the object.

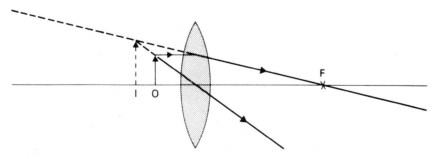

Figure 5.16 Ray diagram

Three facts should be stated about any image:
(1) Real or virtual.
(2) Erect or inverted.
(3) Magnified or diminished.

The linear magnification of an image $= \dfrac{\text{size of image}}{\text{size of object}}$. It may be shown that this equals $\dfrac{\text{image distance}}{\text{object distance}}$.

$$\text{Magnification} = \frac{v}{u}$$

In the two arrangements above:
   Figure 5.15 shows the image as real, inverted, and same size.
   Figure 5.16 shows the image as virtual, erect, and magnified.
By drawing a ray diagram *to scale* the nature of an image may be determined.

Note: In ray diagrams lenses can be represented simply by:

↕   for a converging lens   and   )(for a diverging lens.

Full lenses have been drawn in the above figures to emphasise the lens types.

Images formed by a converging lens

For any converging lens of focal length f:

Object distance, u	Type of image	Examples
(a) $u > 2f$	Real, inverted, diminished	Camera, objective lens of telescope
(b) $u = 2f$	Real, inverted, same size	Terrestrial telescope, photocopier
(c) $f < u < 2f$	Real, inverted, magnified	Projector, objective of microscope
(d) $u = f$	At infinity	Spotlight, eyepiece of telescope
(e) $u < f$	Virtual, erect, magnified	Magnifying glass, eyepiece of microscope and telescope

Images formed by a diverging lens

For *any* diverging lens a virtual, erect, and diminished image is obtained for *all* object positions (Figure 5.18). This type of lens is used in spectacles and certain telescopic arrangements.

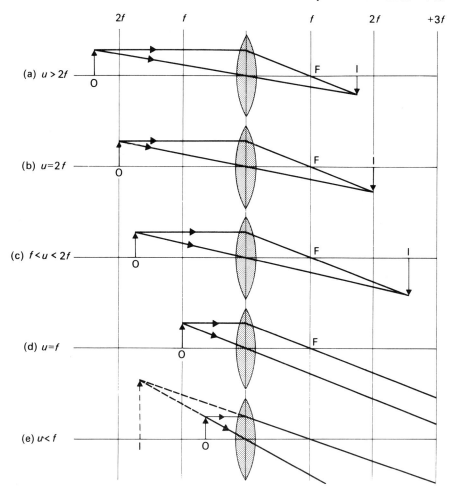

Figure 5.17 Ray diagrams showing the images formed by a converging lens

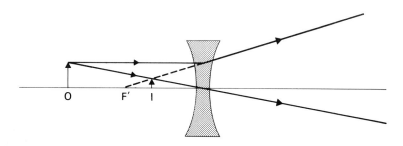

Figure 5.18 Ray diagram showing the image formed by a diverging lens

Note: A lens will have the same effect on light if it is reversed. Thus a ray diagram will remain the same if the *direction* of the incident light is *reversed*, but with the image and object interchanged. For example, if parallel light from an object is incident on a converging lens the image would be formed at the focus, i.e. Figure 5.17(d) with the *direction* of light reversed and the image at the arrow marked O. Observe that light from an object *off* the principal axis will form an image also *off* the principal axis.

Experiment 5.6   *The nature of images formed by lenses*

The nature of **real images** with spherical converging lenses may be shown on a screen, using a simple optical bench. The change in the image when the object distance is altered should be observed. Magnifications should be measured for a range of object distances using a lens of known focal length.

The **virtual images** formed by both types of lenses can be compared by observing the printed page through the lens. Remember that for a virtual image, the object and image are *both* on the same side of the lens. Hence the observer must look *through* the lens at the object in order to see the image. (Figure 5.17(e) and Figure 5.18 illustrate this point.)

Note: The foci and images are *blurred* when wide beams of light are used as only rays near the centre of a lens are brought exactly to the principal focus F or F'. This may be clearly shown with a ray box and five widely spaced incident rays. This effect may also be observed with thick lenses, (see Problem 5.8).

*Lens formula**

Image distances may be calculated using the formula:

$$\frac{1}{f} = \frac{1}{u} + \frac{1}{v}$$

where f is the focal length, u the object distance, and v the image distance.

This formula applies to any lens, converging or diverging, providing a 'real is positive' sign convention is used. This requires that:

(1) All distances are positive if actually traversed by the light, that is distances from real foci, images or objects.
(2) Distances to virtual foci or images are taken to be negative, these distances are only *apparently* traversed by the light. Thus, the focal length of a converging lens is positive and the focal length of a diverging lens is negative.

(There are other sign conventions which will not be discussed here.)

Example: Determine the position of an image of an object placed 30 cm in front of a converging lens of focal length 10 cm.

Using
$$\frac{1}{f} = \frac{1}{u} + \frac{1}{v}$$

then
$$\frac{1}{+10} = \frac{1}{+30} + \frac{1}{v}$$

giving
$$v = +15 \, \text{cm}$$

---

* The lens formula is only for interest or the Special Topics.

The positive sign indicates a real image, situated 15 cm from the lens. The magnification $\frac{v}{u} = \frac{15}{30} = 0.5$, which indicates a diminished image.

Example: Determine the nature of an image of an object situated 20 cm in front of a diverging lens of focal length 10 cm.

Using
$$\frac{1}{f} = \frac{1}{u} + \frac{1}{v}$$

then
$$\frac{1}{-10} = \frac{1}{+20} + \frac{1}{v}$$

giving
$$v = -6.67 \, cm$$

The image is therefore virtual, 6.67 cm from the lens, on the same side as the object, and diminished.

Note: The lens formula enables the image distance, and hence the magnification to be calculated. The sign associated with the image distance indicates if the image is real or virtual but the formula cannot yield information as to whether the image is inverted or erect. A ray diagram, drawn to scale, will give the three points needed to describe an image, which were specified on page 146.

Power of a lens

The power of a lens is defined as the reciprocal of the focal length.

$$\text{Power} = \frac{1}{\text{focal length in metres}}$$

Unit: dioptre (D).

A converging lens has a positive power. A diverging lens has a negative power. Thus a converging lens of focal length 20 cm has a power of $+5 \, D$ but a diverging lens of focal length 25 cm has a power of $-4 \, D$.

Chromatic aberration

A single lens may have **chromatic aberration** because the red light will *not* be brought to the same focus as the blue light. Hence the image of an object may have coloured edges.

This effect may be reduced by using an **achromatic** doublet which consists of a powerful convex lens of low dispersive power with a weak concave lens of high

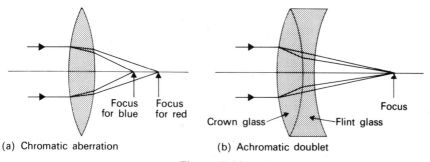

(a) Chromatic aberration   (b) Achromatic doublet

Figure 5.19

dispersive power. The overall effect is a converging lens with little dispersion. Achromatic doublets are used in optical instruments to reduce chromatic aberration.

5:B.2 *Illuminance*

Light provides illumination and makes objects visible to the eye. In the camera the *amount* of light falling on the film is particularly important. In this section only wavelengths of *visible* light will be considered.

A source of light emits energy. The **luminous flux** of a *source* is the energy emitted per second, hence it is a measure of the *power emitted*.

The **illuminance** of a *surface* E is defined as the luminous flux falling on unit area of that surface. Thus illuminance is a measure of the *power received on unit area*.

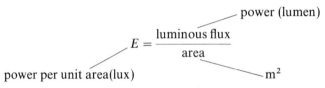

$$E = \frac{\text{luminous flux}}{\text{area}}$$

power (lumen)

power per unit area (lux)

m²

(The SI units of illuminance and luminous flux, the lux and lumen respectively, involve the basic unit the **candela** which is not discussed at this level.* The candela is a measure of the amount of light radiated in a certain direction.)

The illuminance of a surface E used to be called the *intensity of illumination*, a term which can still be encountered in some texts.

Experiment 5.7 *Illuminance and distance from a source of light*

Figure 5.20 Experimental arrangement

The experiment must be carried out in blackout conditions, and care taken to *prevent* reflected light reaching the mirror.

The light meter consists of a photocell and a scale. When light is incident on the sensitive material in the photocell a p.d. is developed, or electrons are emitted, depending on the type of photocell. The magnitude of the p.d. or current depends on the light *received*, that is the flux received over the *fixed* area of sensitive material. Thus the meter records illuminance E, the power incident on this given area. (The meter could be a Philip Harris lightmeter, AVO LM4, or a phototransistor plus milliammeter, but not a photographic light meter.)

The reading on the meter is recorded for different values of the distance d. A graph of light meter reading and $\frac{1}{d^2}$ is plotted which gives a straight line showing that

---

* For interest only, the definition of the candela is given in Section 8:C.2.

$$E \propto \frac{1}{d^2}$$

This experiment shows that the light received on a surface varies inversely with the square of the distance when the surface is moved away from the source. The power from a source will spread out in all directions into a sphere around that source.

5:B.3 *The Camera**

The camera consists of:

(1) A converging **lens** system, which may contain an achromatic doublet and separate lens components.

(2) A **focussing** control to alter the lens-to-film distance.

(3) An **aperture** or **stop** to ensure that the light is incident on the central part of the film and to limit the *amount* of light falling on the film. It takes the form of a circular hole of variable diameter.

(4) An **exposure** control to adjust the *time* for which the shutter is open.

(5) A light sensitive **film** at the back of the camera.

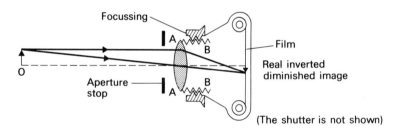

Figure 5.21 The camera

The lens system refracts the light from the subject to form a clear image on the film. '*Focussing*' is obtained by moving the lens which alters the *image distance*. The word 'focussing' here implies 'obtaining a clear image'. When the lens is moved to position B, the image distance is as *small* as possible and usually *equal* to the focal length of the lens. The camera will then bring distant objects into focus on the film. For near objects the lens is moved outwards to position A, (Figure 5.21).

In practice, only certain objects at and around a particular distance will be in focus, the background or objects well in the foreground may be out of focus depending on the setting of the focus control. There is a certain **depth of field**, that is a *range* of object distances, which will give images sufficiently sharp on the film.

The amount of light falling on the film is adjusted by the stop setting and

* Section 5:B.3 is only for interest or the Special Topics.

exposure time controls. These two are interdependent, for instance if the diameter of the aperture is increased, more light enters the camera and the exposure time is reduced.

5:B.4 *The Telescope**

The astronomical telescope

The telescope is a device designed for viewing distant objects. The astronomical telescope has two converging lens systems. The objective lens has a long focal length and the eyepiece a short focal length.

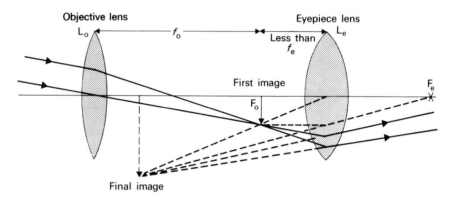

Figure 5.22 Astronomical telescope

Since the object is at a great distance, the first image will be formed at the focus F_O of the objective lens. This first image is real, inverted, and much smaller than the object, but as it is nearer to the eye it *appears* larger. The eyepiece acts as a magnifying glass, hence the distance of the first image (the object for L_e) from the eyepiece must be *less* than the focal length f_e, giving the final image. This final image is virtual, inverted, and diminished with respect to the object, but *nearer* to the eye.

The distance between the lenses (i.e. the length of the telescope) is therefore *less* than the sum of the focal lengths of the two lenses.

Final image at infinity (telescope in normal adjustment)

For a relaxed position of the eye, the final image position is at infinity. The object distance for the eyepiece lens L_e must equal its focal length f_e, hence the first image is now at a distance f_e in front of the eyepiece.

With a final image at infinity, the length of a telescope is *equal* to the sum of the focal lengths of the two lenses, as shown in Figure 5.23.

Magnifying power of a telescope in normal adjustment

The **magnifying power (angular magnification)** *M* is defined as the ratio of the angle subtended by the image, β, to the angle subtended by the object, α.

---

* Section 5: B.4 is only for interest or the Special Topics.

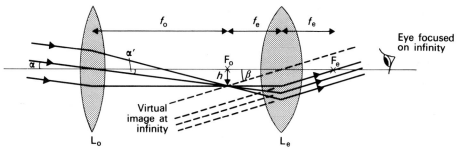

Figure 5.23 Length of a telescope in normal adjustment

Referring to Figure 5.23 and observing that angles α and β are small:

$$\alpha = \alpha' = \frac{h}{f_o}, \quad \beta = \frac{h}{f_e}, \quad \text{and} \quad M = \frac{\beta}{\alpha}$$

Hence

$$M = \frac{f_o \text{ (focal length of objective)}}{f_e \text{ (focal length of eyepiece)}}$$

To increase the magnification of a telescope the focal length of the objective should be made as long as possible and the focal length of the eyepiece very short.

Terrestrial telescope

To convert an astronomical telescope to a terrestrial telescope, an extra converging lens is employed between the objective and eyepiece to invert the image, as shown in Figure 5.24. The final image is now *erect* but still virtual. The distance of this extra lens from its object, that is the first image at F_o, is twice the focal length of this extra lens.

The length of this telescope is $f_o + f_e + 4f$ (erecting lens).

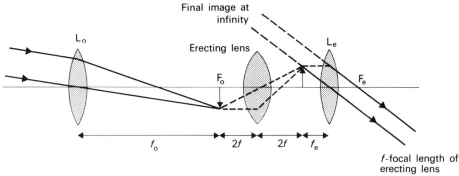

Figure 5.24 Terrestrial telescope

5:B.5   *The Spectrometer*

This is an instrument designed to produce, view, and study spectra. (A simpler version, which is often unsuitable for taking accurate measurements, is termed a **spectroscope**.)

There are three main parts and these are shown in Figure 5.25.

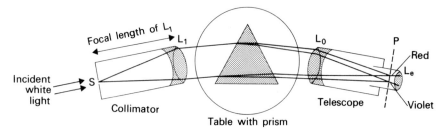

Figure 5.25   Spectrometer

(1) The **collimator**

This has a slit S at one end and a lens L_1 at the other to produce a narrow parallel beam of light.

(2) The **prism** or **diffraction grating**

These split up the incident light into different wavelengths or colours. The important function here is *dispersion* to obtain different colours in different directions.

(3) The **telescope**

Here it is used to view each wavelength clearly. Each wavelength or colour is brought to a separate focus in the focal plane P of the objective L_o and viewed through the eyepiece L_e. (The cross wires are also at the focal plane P.)

The prism or diffraction grating (see page 169) is placed on a central table which is adjusted into a horizontal position by means of levelling screws. This table can be rotated in a horizontal plane in order to bring a diffraction grating into a position perpendicular to the incident light or a prism to a position of minimum deviation. A locking screw ensures no change of position during the experimental investigation.

The telescope is joined by an arm to the central 'leg' of the table. It may rotate about the table which has a scale in degrees on its outer circumference, so that the position of a spectral line may be recorded. In this way the spectrometer may identify elements in a given sample of material by comparison of the spectrum formed from it with the known spectral lines of the pure elements, (see Section 7:A.2).

Problems

5.6   Two convex lenses A and B of the same shape are made of different glass. The refractive indices are 1.50 and 1.65 for A and B respectively. Which has the larger focal length?

5.7   A virtual erect image of an object was obtained. Was a converging or a diverging lens used? Comment on your answer.

5.8 Complete the diagrams in Figure 5.26 to show refraction through the lenses.

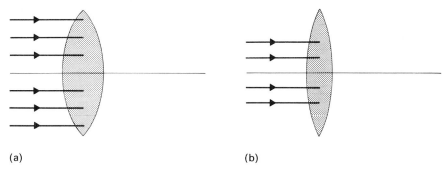

(a)   (b)

Figure 5.26

5.9 Draw a ray diagram to scale to determine the nature of the image of an object situated 15 cm from a converging lens of focal length 10 cm. State the range of object distances, which could be used with this lens in a projector, to produce a magnified image on a screen.

5.10 A transparent slide 3.5 cm wide is placed 15 cm in front of a converging lens of focal length 20 cm. Draw a ray diagram to determine the position of the image. State the width of the image. Would this arrangement be of any use in either a slide projector or a slide viewer?

5.11 A light meter was placed at different distances from a lamp and the following results recorded.

Distance m	1.0	0.8	0.6	0.4	0.3
Light meter reading (lux)	30	45	85	185	365

(*a*) Plot a graph of illuminance (meter reading) and distance.
(*b*) Plot another graph to show the relationship between illuminance and distance.

C: Waves and Wave Optics

5:C.1 *Basic Wave Properties*

When a 'source' vibrates about a mean position a disturbance, that is a wave, may travel away from the vibrating source. The particles along the path of the wave also tend to vibrate.

Example: A water wave can be generated by placing a vibrating pin, the 'source', in the surface of the water. The individual particles of water will vibrate up and down in a vertical direction but the wave travels horizontally across the surface of the water causing the particles of water in its path to vibrate, as shown in Figure 5.27(a).

Example: A sound wave may be generated by a tuning fork which alternately compresses and rarefies the air adjacent to the prong. Each layer of air will vibrate about a mean position in a horizontal direction while the wave travels away from the source also

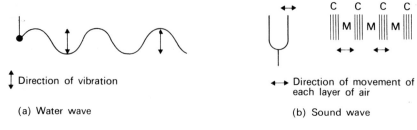

↕ Direction of vibration

(a) Water wave

⟷ Direction of movement of each layer of air

(b) Sound wave

Direction of wave propagation ⟶

Figure 5.27 Waves

in the horizontal direction, as shown in Figure 5.27(b). In this figure C denotes a compression while M denotes a rarefaction. At a slightly later time, of course, the high pressure will have moved outwards to the position at present marked M. The value of the pressure also varies periodically in the horizontal direction.

A **transverse wave** is one in which the vibration of the particles is perpendicular to the direction of propagation of the wave. For example: a water wave, a vibrating string or wire, electromagnetic radiation.

A **longitudinal wave** is one in which the vibration of the particles is in the same direction as the wave propagation. For example: sound waves, a slinky spring.

Some basic terms

For both types of waves the figure below shows the instantaneous magnitude of the disturbance x against the distance. x may be the distance moved vertically by water particles, the pressure variation of air in sound waves or the magnitude of the electric field in light waves.

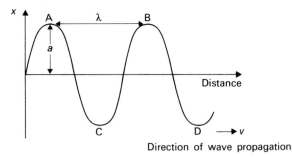

Figure 5.28 Graphical representation of a wave

λ The **wavelength** is the distance between two consecutive maxima or minima positions, that is the distance A → B or C → D. (Unit: metre, m)

f The **frequency** is the number of complete waves passing any point in one second. (Unit: hertz, Hz)

T The **period** is the time taken for one maximum to move to the position of the next maximum, that is the time taken for the wave to travel from A to B. (Unit: second, s)

v The **velocity** is the rate at which the wave travels. (Unit: $m s^{-1}$)

a The **amplitude** is the maximum variation of the quantity *x* of the vibrating particles from their mean or rest values. The unit of the amplitude depends on the chosen quantity *x*. For example, the distance in metres from the top of a crest to the undisturbed surface in a water wave, or the maximum excess pressure in pascals of air in a sound wave.

Phase: Two waves of the same frequency and speed are **in phase** if their maxima coincide.

Relationship between v, λ, and f

Since the distance AB in Figure 5.28, is the wavelength and the time taken for a maximum at A to travel to B is the period *T*, then the speed $v = \lambda/T$. However, the frequency *f*, the number of waves per second, is the reciprocal of the period *T*. Hence:

$$f = \frac{1}{T} \quad \text{and} \quad v = \lambda f$$

Interference

If two waves of the same frequency and amplitude are brought together in such a way that their maxima coincide, that is they are *in phase*, the resultant wave will have the same frequency and wavelength but an increased amplitude, Figure 5.29(a).

This is termed **constructive interference**.

Example: A resultant water wave with higher crests; a louder sound note of the same pitch.

(a) Constructive interference   (b) Destructive interference

Figure 5.29 Interference

If, however, the maxima of one wave coincide with the minima of the other then a null resultant is obtained, Figure 5.29(b).

This is termed **destructive interference**.

Examples: A lack of movement on a water surface; silence from two sound waves.

To prove the existence of a wave it is necessary to demonstrate destructive interference.

Diffraction

Diffraction may be defined as a *bending* phenomena; that is the tendency for waves to 'go round corners' giving wave motion in the geometrical shadow of an obstacle or aperture.

(1) For a slit or aperture this diffraction effect is greater when the *aperture* is *small*. The greatest effect is obtained when the width of the aperture approaches the wavelength of the waves, Figures 5.30(a) and (b).

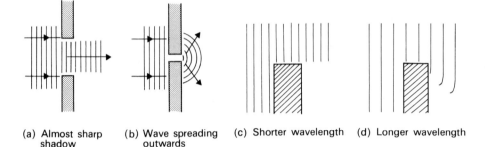

(a) Almost sharp shadow   (b) Wave spreading outwards   (c) Shorter wavelength   (d) Longer wavelength

Figure 5.30   Diffraction through apertures and around obstacles

Note: The direction of wave propagation is perpendicular to the wavefronts.

(2) When waves are incident on a single sharp edge the *longer* wavelength wave is bent to a greater extent into the geometrical shadow, Figure 5.30(c) and (d).

(3) Diffraction effects are also observed when waves are incident on an obstacle. Notice that for a large obstacle, as in Figure 5.31(a), almost no diffraction occurs and a sharp shadow is obtained. For small obstacles the wavefront reforms beyond the object, Figure 5.31(b), but interference of the individual segments of the waves will give positions of maxima and minima.

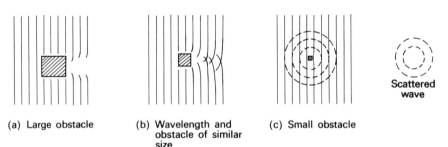

Scattered wave

(a) Large obstacle   (b) Wavelength and obstacle of similar size   (c) Small obstacle

Figure 5.31   Diffraction of light waves

Scattering

For a very small object, about the size of the wavelength, some scattering of the wave can occur. The object can act as a new source of wavefronts by absorbing and re-radiating the incident energy. This gives rise to a superimposed circular pattern, Figure 5.31(c). It is these phenomena which account for the blue colour of the sky. The blue light with the shorter wavelength is scattered more by the small atmospheric particles than red.

Diffraction of light waves

Diffraction effects are more noticeable for narrow gaps and long wavelengths. Thus a very fine slit and red light are required to show light diffraction. The diffraction pattern for red and blue light incident on a *narrow* slit is shown in Figure 5.32. Observe the positions of minima on the diffraction pattern caused by destructive interference of the overlapping segments of the wave. These dark and bright bands are called *fringes*.

(a) Red light   (b) Blue light

Figure 5.32   Diffraction of light at a single slit

Reflection

If waves are incident on a large opaque barrier they will be reflected.

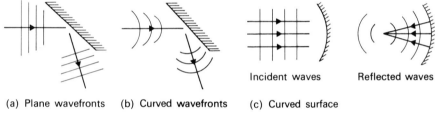

(a) Plane wavefronts   (b) Curved wavefronts   (c) Curved surface

Figure 5.33   Reflection

The wavelength, frequency, and velocity will *not* change as the incident and reflected waves are in the same medium, but the curvature of the wavefronts alters except in the case of plane waves incident on a plane barrier.

Note: There is a phase change of half a cycle on reflection.

Standing waves

When a wave is reflected back along the incident path interference may take place between the incident and reflected waves yielding **standing waves**.

Example: A vibrating wire, fixed at one end, F.
   No movement is observed *along* the wire as the wave is **stationary**. The only movement is the vibration of segments of the wire in the vertical direction.

The overall picture in Figure 5.34(b), of the instantaneous positions of the wire, shows that certain points N, called **nodes**, are stationary while other points A, called **antinodes**, vibrate from a maximum in one direction to a maximum in the other. All the intermediate points along the wire vibrate with amplitudes less than the maximum of the antinodes.

Note: For standing waves, all points along the wave reach their *different* amplitudes at the *same* time. In a travelling wave each point along the wave reaches the same amplitude but at different times.

For standing waves the distance between two consecutive maxima, antinodes, is $\frac{1}{2}\lambda$, and the distance between consecutive nodes is also $\frac{1}{2}\lambda$, where λ is the incident wavelength.

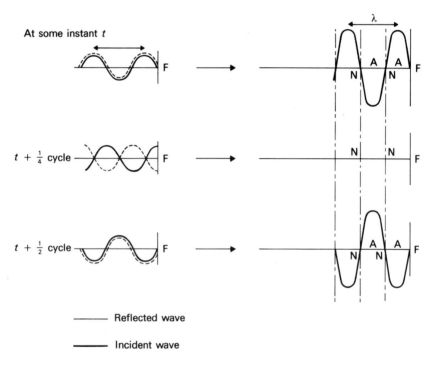

Reflected wave

Incident wave

(a) Formation of standing waves

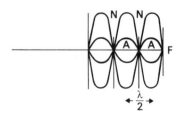

(b) Overall picture with intermediate positions

Figure 5.34 Standing wave

Refraction

Refraction is the change in *direction* of a wave when it passes obliquely from one medium to another. This change in direction arises because the *velocity* of a wave *alters* when it passes into different media.

The precise change in velocity depends on the type of wave as well as the media concerned. For example, light waves have their maximum velocity in a vacuum and travel faster in gases than solids; whereas sound waves travel slower in gases than in solids. The table below indicates velocity changes for different situations.

Change in velocity trends for different media

Type of wave	Higher velocity	⟶	Lower velocity
Water waves	deep water		shallow water
Light	vacuum air	water	glass
Sound	solids	liquids	gases

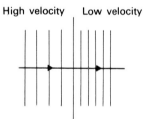

(a) Incorrect!

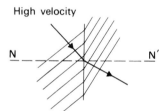

(b) Change in direction

High velocity   Low velocity

(c) No change in direction

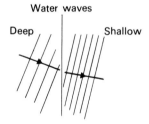

(d) Shallow water waves have the lower velocity

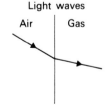

(e) Refraction of light waves

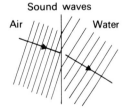

(f) Sound waves travel faster in water

Figure 5.35 Refraction

The frequency remains *constant* when waves are refracted from one medium to another; the frequency depending only on the number of waves emitted per second by the *source*. Remembering that $v = \lambda f$, if the velocity v changes the wavelength λ will change in the same proportion, the frequency f remaining constant see Section 5:A.5 on page 144.

Hence when waves pass from a high velocity medium to a low velocity medium their wavelength will decrease. Because no extra waves are generated at the interface, Figure 5.35(a) must be incorrect! This implies that a change of direction must occur so that the wavefronts are continuous across the interface as shown in Figure 5.35(b).

As mentioned previously the direction of travel of a wave is perpendicular to the wavefronts, thus Figure 5.35(b) demonstrates that when a wave is refracted from a medium of high velocity to a medium of low velocity the wave is bent towards the normal. The **normal** NN′ is a line drawn perpendicular to the interface. For plane waves incident perpendicular to the interface, as in Figure 5.35(c), each wavefront is in one medium only hence no change of direction occurs. Compare the change of direction occurring when light enters glass, Figure 5.35(e), with that of sound entering water, Figure 5.35(f).

Note: The frequency of waves remains constant during refraction but the velocity and therefore the wavelength alter. However, if the frequency of the *source* is altered the refraction, that is the precise change in direction, will be different.

Experiment 5.8

A ripple tank may be used to demonstrate the wave phenomena in this section. Water waves are particularly useful for illustrating wave phenomena since the wavefronts can be seen and their wavelengths are a convenient size of a few cm. Sound waves are invisible, and those pleasing to the ear have long wavelengths. The wavelengths of light are very small, but those of microwaves are around 3 cm.

5:C.2 *Interference of Electromagnetic Radiation—Quantitative*

Path difference and interference

Two sources S_1 and S_2 emit waves of the same frequency, wavelength, and phase. For microwaves or light, S_1 and S_2 would be two slits in a barrier placed in front of a *single* source.

The two sources will interfere. If, at any position, a 'crest' from S_1 coincides with a 'crest' from S_2, a maximum will be obtained. But if a 'crest' coincides with a 'trough' there will be a minimum.

Thus at any point P, for a **maximum**, the *path difference* between PS_1 and PS_2 must be zero or a whole number of wavelengths, in order that the two waves arrive *in phase*.

For a maximum at P the path difference is zero or an integral number of wavelengths.

$$PS_2 - PS_1 = n\lambda \qquad n \text{ an integer } 0,1,2\dots.$$

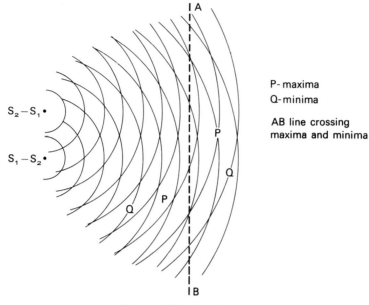

Figure 5.36 Interference

For a **minimum** at a point Q, the path lengths QS_1 and QS_2 must be such that the waves arrive at Q half a cycle out of phase, i.e. the *path difference* contains an odd half wavelength.

For a minimum at Q the path difference is an odd number of half wavelengths.

$$QS_2 - QS_1 = (n + \tfrac{1}{2})\lambda \qquad n \text{ an integer}$$

These two relationships can be combined into a *single* equation (see Problem 5.17 page 176).

Moving along the line AB in Figure 5.36, positions of maxima and minima would be passed. If S_1 and S_2 were red light waves these would be positions of dark and red light, (or if sound waves then positions of loud and quiet).

Experiment 5.9 *Interference of microwaves*

Microwaves have much longer wavelengths than light waves. Interference effects can be demonstrated using a double slit separation of a few cm. A probe detector is moved to a position Z which records a radiation maximum and the distances of Z to S_1 and S_2 measured. The path difference $ZS_2 - ZS_1$ is calculated. The probe is then moved to other positions and both the path difference, and the intensity of the radiation received by the detector noted. It is observed that the radiation at any point depends on the path difference, not just on the distance from the sources.

The **wavelength** of the microwaves can be determined by positioning the transmitter in front of a single barrier to *reflect* the waves back along their incident path. Standing waves are then produced. Using the probe detector, the distance across five minima can be determined. Since the distance between consecutive minima for standing waves is $\lambda/2$, the wavelength can be calculated.

Using this wavelength and the above path lengths the relationships for a maximum or a minimum position can be verified.

Example: A double slit, of separation 5 cm, is positioned in front of a 3 cm microwave source.

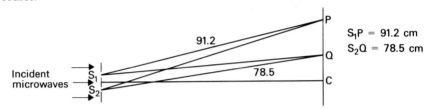

Figure 5.37   Interference of microwaves

A detector is moved slowly away from position C and passes a minimum at Q, then a maximum at P. State the lengths of S_1Q and S_2P.

At Q, the first minimum

$$S_2Q - S_1Q = \frac{\lambda}{2}$$

giving

$$S_1Q = 78.5 - 1.5 = 77.0 \, \text{cm}$$

At P, the first maximum

$$S_2P - S_1P = \lambda$$

giving

$$S_2P = 94.2 \, \text{cm}$$

(It should be mentioned that in this arrangement CP has a value of 55.6 cm which indicates that the angle $P\hat{M}C$ is *not* small. Compare this with the angle θ of the Young's slits experiment below.)

Coherent and incoherent waves

Light is produced from excited atoms returning to their ground state (see Section 7:A.2), a finite wave train of light being emitted by each atom. The light from one source, of excited neon atoms, for example, will not produce waves of identical amplitude and phase as light from another source of neon atoms, although both sources generate the same wavelengths. The waves from two such sources are said to be **incoherent** even though their frequencies, wavelengths, and average amplitudes may be the same. To obtain two **coherent** beams of light, that is beams which have identical amplitudes and phase at every instant, a *single* source must be used and divided into two beams.

Interference of two coherent light sources — Young's slits

Experiment 5.10   *Young's slits experiment*

Two sets of coherent waves are produced by passing a narrow beam of light through two small slits, as shown in Figure 5.38.

Two narrow 'slits' are scratched on a blackened microscope slide. The width of each slit must be narrow as the wavelength of light is small. If red light is used, a series of black lines on a red background is observed. The black lines show positions of destructive interference.

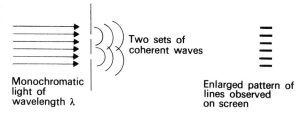

Figure 5.38 Young's slits experiment

*Theory—Relationship between fringe separation and wavelength**

For a bright fringe on the screen the path difference between $S_2 P$ and $S_1 P$ in Figure 5.39 must be an integral number of wavelengths. If P is the first bright fringe then

$$S_2 P - S_1 P = S_2 A = \lambda$$

(A is a point on PS_2 such that $PA = PS_1$.)

Because PM is large compared to $S_1 S_2$, PM intersects $S_1 A$ almost at a right angle, i.e. $M\hat{Q}S_1 = 90°$, giving $M\hat{S}_1 Q = \theta$.

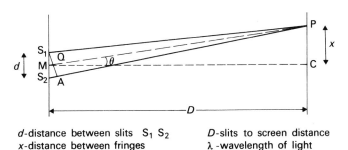

d-distance between slits $S_1 S_2$ *D*-slits to screen distance
x-distance between fringes λ -wavelength of light

Figure 5.39

Since D is very much larger than d or x, θ is a small angle hence

$$\sin \theta = \tan \theta = \theta.$$

$$\theta = \frac{\lambda}{d} \quad \text{(from } \Delta S_1 A S_2 \text{)}$$

$$\theta = \frac{x}{D} \quad \text{(from } \Delta PMC \text{)}$$

Giving

$$\frac{\lambda}{d} = \frac{x}{D}$$

and

$$\lambda = \frac{xd}{D}$$

Note: The interference of the two waves gives constructive interference, a bright band, when the path difference is λ or an integral multiple of λ. Hence a

* This proof is not required for examination purposes.

bright fringe is obtained at a position x, $2x$, $3x$,..., nx where n is an integer. Similarly, a dark fringe is observed for a path difference of an odd number of *half* wavelengths. Thus for a fringe *separation* of Δx

$$\lambda = \frac{\Delta x d}{D}$$

where Δx is the separation between two consecutive maxima *or* two consecutive minima.

For a given wavelength the distance apart of the fringes may be increased by reducing the distance between the slits d or by increasing the distance to the screen D. If light of a shorter wavelength is used, say blue light, the path difference for the first bright fringe at P must still be one wavelength, hence P will be nearer to C, that is x is decreased. This may be directly deduced from the formula $\lambda = \frac{\Delta x d}{D}$ if $\lambda \downarrow$ then $\Delta x \downarrow$.

Example: A monochromatic beam of light is passed through a double slit of separation 0.25 mm situated 3 m away from a screen. The distance across 6 minima was measured as 4.8 cm. What colour of light was used?

Using
$$\lambda = \frac{\Delta x d}{D} \qquad \Delta x = \frac{4.8}{6} = 0.8 \,\text{cm}$$

$$\lambda = \frac{0.8 \times 10^{-2} \times 0.25 \times 10^{-3}}{3} \qquad \text{all lengths in m}$$

$$\lambda = 6.67 \times 10^{-7} \,\text{m}$$

The wavelength is 667 nm which is red light.

When a white light source is employed fringes will be obtained at different positions for each wavelength. A central bright patch at C, where the path difference is zero for all wavelengths, will have a series of overlapping spectra on either side; the blue, with the shortest wavelength, requiring the least path difference, being nearest to C.

Experiment 5.11 *Measurement of the wavelength of light*

The Young's slits arrangement, described above, is used. If a laser is available, the intensity is sufficient to enable a good interference pattern to be obtained on a screen. If a laser is not available, a single slit is placed in front of a source of red light. A scale drawn on a piece of paper is placed just above the source. The observer, standing 3 or 4 metres away, must look through the double slit held in the hand to observe the fringes by the scale, and estimate Δx. A more accurate method is to place the double slit at one end of a long cardboard tube and take a photograph of the interference pattern on a film placed at the other end. The distance across a number of fringes can be measured to obtain Δx. The slit separation can be measured with a travelling microscope, but the uncertainty will be rather large. The slit-to-fringe separation is measured directly. Using $\lambda = \frac{\Delta x d}{D}$ the wavelength is calculated.

5:C.3 *The Diffraction Grating*

A diffraction grating has a large number of equally spaced slits, probably several thousand per centimetre. Interference takes place between all the coherent waves passing through these slits.

Directly opposite the grating there will be a *central* maximum, since the path length is the *same* for all the waves.

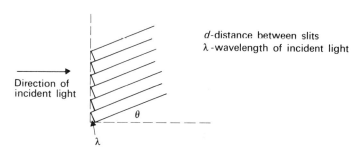

d-distance between slits
λ-wavelength of incident light

Direction of incident light

θ

λ

Figure 5.40 Diffraction grating

However, at some angle θ each wave is λ ahead of the next. If all these waves are brought together by a convex lens then a bright resultant will be formed by constructive interference. At other angles, each wave will destructively interfere with some other wave giving darkness. A bright fringe will be obtained in a direction θ when

$$d \sin \theta = \lambda$$

Another bright line will be obtained for a larger angle θ' when $d \sin \theta' = 2\lambda$, each wave now being *two* wavelengths ahead of the next.

If white light is used then each wavelength $\lambda_1, \lambda_2, \lambda_3, \ldots$, that is each colour, will give a bright fringe in a slightly different direction $\theta_1, \theta_2, \theta_3, \ldots$, such that $d \sin \theta_1 = \lambda_1$, $d \sin \theta_2 = \lambda_2, \ldots$, for each colour. Thus the incident light will be split up into a spectrum by the grating. Remember that the blue with the shorter wavelength and smaller value of θ, will be nearest the central white patch. This spectrum is called the **first order spectrum**. At larger angles θ', such that $d \sin \theta' = 2\lambda$ for each wavelength, another spectrum will be obtained called the **second order spectrum**. Higher order spectra are less likely as $\sin \theta$ must be less than unity! Also, if they do occur, they will tend to overlap.

Thus in general, for each wavelength λ the nth order maximum is given by:

$$d \sin \theta_n = n\lambda$$

Comparison of the spectra produced by a prism and a diffraction grating

A **prism** produces a *single* spectrum. It is caused by the difference in *refraction* of the various colours, red through to violet. To increase dispersion a material of high dispersive power is required.

The **diffraction grating** produces a *series* of spectra, on both sides of the central white patch. These are caused by *interference* of the waves from each slit. To increase the dispersion, the grating spacing *d* must be reduced.

A diffraction grating has an advantage over a prism in that little light is lost by reflection or absorption.

A **double slit** also produces a series of spectra on both sides of a central maximum. However these spectra tend to be very close together and over-lapping.

Example: Calculate the fringe separation for blue, green, and red light of wavelengths 400 nm, 540 nm, and 680 nm respectively, if the light is passed through a double slit of separation 0.2 mm and observed from 3 m away. Comment on the clarity of any spectrum produced.

Using $\Delta x = \dfrac{\lambda D}{d}$ for each wavelength gives:

$$\Delta x \,(\text{blue}) = 6\,\text{mm} \qquad \Delta x \,(\text{green}) = 8.1\,\text{mm} \qquad \Delta x \,(\text{red}) = 10.2\,\text{mm}$$

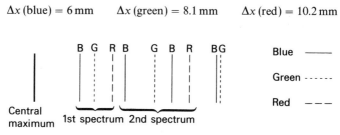

Figure 5.41

Only spectra on one side of the central maximum are shown. The first spectrum will be observed, but notice that the blue of the third spectrum occurs before the red of the second. Thus only one spectrum on each side could be clearly observed without overlapping colours. In practice it may be difficult to discern even this one spectrum.

Note: The effects produced by a prism, single slit, double slit, and diffraction grating, using a monochromatic source, should be compared in terms of the patterns produced, (see Problem 5.16).

5:C.4 *Electromagnetic Radiation*

Light is electromagnetic radiation of certain wavelengths. The full range of electromagnetic radiation is given in Figure 5.42.

Note: A logarithmic scale has been used in Figure 5.42.

All electromagnetic radiation has a velocity of $2.998 \times 10^8 \,\text{m s}^{-1}$ in a vacuum and lower velocities in other media. They show refraction, reflection, inter-ference, and diffraction similar to other wave motions, but different wavelengths exhibit different degrees and effects in different media. Microwaves and light do not have the same refraction in water; X-rays and infra-red will not show interference in the same apparatus as their wavelengths are not the same!

Previously it has been mentioned that electromagnetic wave motion is a transverse wave. The physical quantity or 'particle' which is varying will now be considered. In water waves the water vibrates up and down, in sound the excess air pressure periodically varies.

An electric charge has an associated electric field E. A moving charge, or charges, which constitutes an electric current will, in addition, have an associated

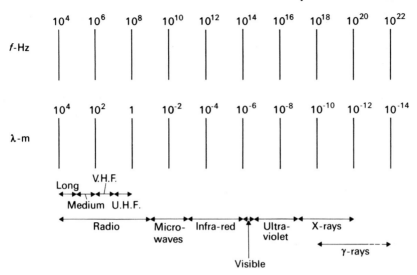

Figure 5.42 The electromagnetic spectrum

magnetic field *B*. Charges moving with changing velocities, that is accelerating charges, will have *changing* magnetic and electric fields.

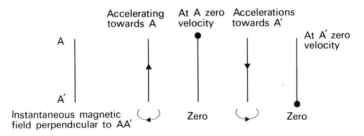

Figure 5.43 Magnetic field around an oscillating electron

If a single electron oscillates in a wire AA′ (Figure 5.43) the magnetic field varies from a maximum in one direction to a maximum in the other.

The overall picture of the variation in the magnetic field with time, as it spreads away from AA′, is shown in Figure 5.44.

Note: The direction of the magnetic field is perpendicular to the page, that is perpendicular to AA′.

Figure 5.44 Variation of the magnetic field

As the electron is accelerating the electric field will also vary periodically but in a plane perpendicular to the magnetic field (Figure 5.45).

The electromagnetic wave will radiate outwards in all directions from the wire AA' but perpendicular to AA'.

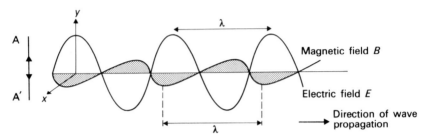

Figure 5.45 Variation of E and B

Note: In electromagnetic radiation *two* physical quantities periodically vary, namely the magnetic field and the electric field. These fields E and B are perpendicular to each other and they are both at right angles to the direction of the wave propagation.

The frequency of the wave will depend on the frequency of oscillation of the electrons in the wire.

The electromagnetic spectrum

γ-**rays** may be emitted when excited **nuclei** return to a lower energy state (see Section 7:B.2). They are observed in radioactive decay and cosmic radiation.

X-rays may be generated when high velocity electrons are suddenly decelerated at a suitable target.

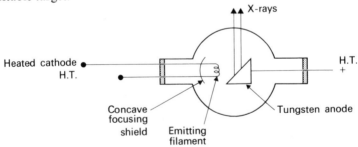

Figure 5.46 *X-ray tube*

A high voltage, of the order of 300 kV, is applied between the cathode and anode in Figure 5.46 to accelerate the electrons and the anode is surrounded by a cooling system.

Uses:

(1) To produce ionization.
(2) In medicine and dentistry to photograph bones, teeth, and other organs.

(3) To measure crystal lattice spacing by diffraction, as the spacings are of the same order of magnitude as the wavelengths of X-rays.

When a diffraction grating is illuminated by light, dark and bright bands are observed. Because a crystal lattice has a regular array of nuclei in three dimensions, it acts like a series of diffraction gratings giving a pattern of spots, not lines. However, it is more usual to examine the diffraction pattern produced by the *reflection* of X-rays from the nuclei in the crystal planes. From examination of the diffraction pattern the lattice spacing may be determined.

Ultra-violet radiation may be emitted when certain excited atoms return to their ground state (see Section 7:A.1).

Uses:

(1) Produces vitamin D in the skin, but is harmful in large doses.
(2) Produces **fluorescence** effects. The fluorescent material absorbs the ultra-violet, then decays in microseconds to a lower energy state, thence to the ground state with the emission of radiation of a lower frequency, namely visible light. In some substances the decay to the intermediate state may take from seconds to hours, and light may be emitted after the ultra-violet has been removed, this latter phenomena is termed **phosphorescence**.
(3) Causes ionization of atoms in the atmosphere producing the ionosphere.

Visible light may also be produced by excited atoms returning to their ground state. Visible light is the relatively *small* range of wavelengths which can be detected by the human eye.

Infra-red radiation is usually emitted by excited molecules returning to lower energy states. This is a heat radiation.

Uses:

(1) Transmission of heat from the Sun, electric fires or radiators.
(2) In military and haze photography, as infra-red radiation penetrates cloud and darkness to a certain extent.
(3) In investigations of molecular structure.

Microwaves may be produced by a klystron or a magnetron, which are special types of valves causing electronic oscillations.

Uses:

(1) In radar when a short burst of microwaves may be reflected back by a metal object. The emitted and reflected waves are rectified and applied to the Y-plates of a c.r.o. where each wave will cause a deflection. From the interval between deflections the distance of the object may be determined.

Example: A burst of microwaves is deflected from an aircraft. A spot on a c.r.o. takes $\frac{1}{100}$ milliseconds to move 1 cm. A deflection due to the emitted wave occurs at the 0.5 cm mark and a second, due to the reflected waves, occurs at the 2.5 cm mark. How far away is the aircraft?

Distance on c.r.o. between deflections $= 2\,\text{cm}$
Time taken for the spot to move $2\,\text{cm} = 10^{-5} \times 2\,\text{s}$
Velocity of microwaves $= 3 \times 10^8\,\text{m\,s}^{-1}$
Distance travelled by microwaves $= 3 \times 10^8 \times 2 \times 10^{-5}\,\text{m}$
$= 6 \times 10^3\,\text{m}$
Distance to aircraft $= 3 \times 10^3\,\text{m}$

Alternatively, a continuous beam of microwaves may be emitted from a rotating parabolic mirror and used to scan the surrounding area. For example, the mirror may be attached to the mast of a ship. Reflected waves from other ships or land masses may be detected and displayed on a screen. Microwaves have the advantage of penetrating fog and cloud.
(2) For determining molecular structure.

Radio waves cover a large range of wavelengths. They may be produced by the oscillations of electrons in an aerial. Their main use is in communications.

Use of radio waves in communications

Radio waves have the important property of being refracted by the ionosphere, a layer of ions formed by the passage of ultra-violet radiation from the Sun. The waves will only be refracted back to the Earth if they are incident on the ionosphere at a certain angle, otherwise they will be lost (ray C in Figure 5.47).

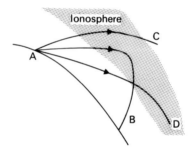

Figure 5.47

Rays emitted in an acute direction D will have a long path length and thus return weaker. The radio waves are usually beamed at a particular angle and secondary transmission stations pick up and reradiate them.

Very high frequency radio waves are not sufficiently refracted by the ionosphere, which has different refractive indices for different frequencies, hence these radio waves, UHF, may be reflected back by artificial satellites.

Radio waves are required to transmit information, but audio frequencies are much lower than radio frequencies, so the audio frequency information is 'carried' by the radio waves. A property of the carrier wave, the amplitude or frequency is varied at the audio frequency, as in Figure 5.48. This is **modulation**.

In *amplitude modulation* the amplitude of the radio wave, the 'carrier wave', is altered to correspond to that of the low-frequency audio wave. In *frequency modulations* the frequency of the radio wave is varied. A radio receiver picks up the radio carrier wave and by demodulation obtains the audio frequency information.

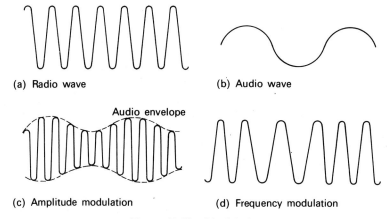

(a) Radio wave

(b) Audio wave

Audio envelope

(c) Amplitude modulation

(d) Frequency modulation

Figure 5.48 Modulation

Problems

5.12 What is the distance between a node and neighbouring antinode of a standing wave in terms of the incident wavelength?

5.13 In a ripple tank the wavelength is observed to alter from 5 cm at position A to 3 cm at position B. What might cause this change? What is the relative change in velocity and frequency from A to B?

5.14 An ultrasonic wave and a light wave each have a wavelength of 500 nm in air. Calculate the frequency of each wave.

5.15 The generators SG in Figure 5.49 are tuned to a wavelength of 0.1 m and are connected to loudspeakers A, B, and C as shown. A microphone M is connected to a c.r.o. which shows the strength of the signal at the position of the microphone. Comment on the distances between two neighbouring minima positions in (a) and (b).

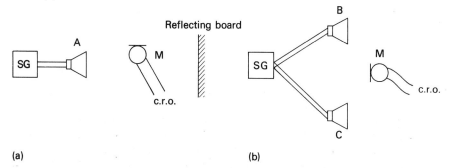

Reflecting board

(a)

(b)

Figure 5.49

5.16 A narrow beam of red light is incident, in turn, on: a 60° prism at minimum deviation, a double slit, and a diffraction grating.

(*a*) Show the path of light through the prism and the patterns produced by the slits and grating.

(*b*) Describe the change in this path or these patterns when the red light is replaced by blue light.

5.17 A source of microwaves is incident on a double slit. At a position X there is a maximum or a minimum when the path difference from X to each slit is given by

$$\text{path difference} = \frac{n\lambda}{2}$$

where n is an integer and the path difference is $XS_2 - XS_1$.

Figure 5.50

(a) Explain what values of n produce a maximum at X.
(b) State what values of n give rise to a minimum.
(c) Where in the sketch must X lie for n to equal zero?
(d) The point P is a maximum. There are 8 complete waves between P and S_1 but only 5 between P and S_2. If the distance $S_2P = 22$ cm calculate the wavelength and the distance S_1P.

5.18 A beam of microwaves of wavelength 2 cm is incident on a double slit. A detector is positioned 30 cm (equidistant) from both slits. The detector is moved round, maintaining a distance of 30 cm from S_1 but increasing its distance from S_2.
(a) State the distance across 8 minima.
(b) How far from S_2 is the eighth minimum position?

5.19 A beam of yellow light, of wavelength 580 nm, from a single slit falls on a double slit and an interference pattern is recorded on a film 2.5 m away. If the double slit separation is 0.2 mm, calculate the distance across four fringes.

5.20 In a Young's slits experiment the distance Δx between fringes was measured in order to determine the wavelength λ used, from the formula $\lambda = \dfrac{\Delta x d}{D}$. State what D and d are.
(a) Considering each quantity Δx, d, and D, comment on the effect of their accuracy of measurement on the accuracy of determining λ.
    What is the effect on the fringes of
(b) altering the distance D,
(c) altering the width of the slits,
(d) changing from yellow to blue light,
(e) using yellow and blue light together?

5.21 (a) Which type of electromagnetic radiation would you use to demonstrate diffraction with a 3 cm aperture?
(b) A clown's hat irradiated with ultra-violet light was found to shine in the dark after the ultra-violet was switched off. What is the term for this effect?

5.22 A parabolic rotating mirror emits microwaves of wavelength 4 cm. The receiving apparatus shows that reflection from an obstacle takes 10^{-5} seconds. What difference would be observed if the wavelength of the waves was increased to 8 cm?

6   Heat and a Particle Model

A: Temperature and Heat

6:A.1   *Temperature*

Unit: kelvin (K), scalar.
Temperature is a measure of the degree of hotness or coldness. Notice that temperature is not a measure of heat; but if two objects are in contact heat will flow from the object of higher temperature to that of lower temperature.

A **temperature scale** may be established by:

(1) choosing a physical property of a certain substance, called a **thermometric property**, which varies with temperature. For example, the expansion of a liquid column of mercury.
(2) selecting two fixed points and measuring the thermometric property at those points. For example, the length of the liquid column in melting ice, the ice point, and in steam from boiling water under normal pressure, the steam point.
(3) designating values to the two fixed points and dividing the interval between them into equal divisions. For example, allot the value 0 to the ice point and 100 to the steam point.

In this way a number of temperature scales can be set up depending on the thermometric property chosen.

Thermometric properties include:

(1) the length of a column of liquid in a glass tube, for example, mercury or alcohol,
(2) the pressure of gas at constant volume,
(3) the resistance of a coil of platinum wire, or a thermistor,
(4) the e.m.f. of a thermocouple. For example, if each end of a piece of constantan wire is joined to a copper wire, and the junctions are maintained at different temperatures, an e.m.f. will be developed which depends on this difference in temperature, see Figure 6.1,
(5) the colour of light emitted by a hot source (useful at high temperatures),
(6) the speed of sound through helium (useful at low temperatures).

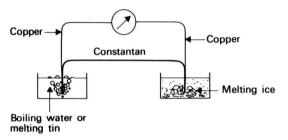

Figure 6.1 A thermocouple

A **thermometer**, an instrument designed to measure temperatures, utilizes a particular thermometric property of a given substance.

Note: Because the various thermometric properties depend in different ways on the temperature, the 'degree of hotness' of a given object will not necessarily have the same value on the different temperature scales of the various thermometers unless that value coincides with a fixed point.

The **constant volume gas thermometer** is chosen as a practical standard and its scale is called the **real gas scale**. The thermometric property of this thermometer is the *pressure* of a fixed mass of gas maintained at a constant volume. Other thermometers may be either calibrated against this thermometer or the temperatures recorded on them corrected to the real gas scale.

For interest only, it may be mentioned that the kelvin, a basic SI unit, can be defined in terms of the two fixed points of absolute zero, $0\,K$, and the triple point of water on the ideal gas scale, whose thermometric property is the pressure of a fixed mass of **ideal gas** (see Section 6:B.4) maintained at constant volume.

The two common fixed points, the ice point and the steam point, on the Celsius scale are $0°C$ and $100°C$, whereas on the kelvin scale these points are $273.15\,K$ and $373.15\,K$ respectively. Hence, to a good approximation, 273 can be added to a temperature θ on the Celsius scale to give the corresponding temperature T on the kelvin scale.

$$\theta + 273 = T$$

Note: An interval of one degree Celsius is the same as an interval of one kelvin, but care should be taken to use temperatures in kelvin in most calculations.

Measurement of temperature—choice of thermometer

When measuring temperatures the choice of thermometer will depend on the range of temperatures involved, the accuracy required, and convenience of measurement. The constant volume gas thermometer may be used over a wide range of temperatures, $\sim 5\,K{-}1750\,K$, but it is an inconvenient thermometer to use. The platinum resistance thermometer, with a range of around $80\,K{-}1400\,K$, and the thermocouple, with a range of $\sim 25\,K{-}1700\,K$, are more useful. Although the mercury-in-glass thermometer has a much smaller range, $-39°C$ to $357°C$, and is not so accurate, it is a useful thermometer for laboratory experiments as it gives direct readings, and is easy to use and is portable.

Note: A thermometer *absorbs* heat from the substance whose temperature it is measuring, thus recording a slightly lower temperature, assuming that the temperature of the substance is above that of the thermometer. It may also take *time* for the thermometer to reach thermal equilibrium with the substance being measured. The thermocouple has the advantage of measuring accurately both local and varying temperatures.

6:A.2 *Heat*

Heat is a form of energy.

There are three distinct states of matter: solid, liquid, and vapour or gas. Some substances sublime when heated, that is they change directly from a solid to a vapour, for example, iodine.

When a solid, a liquid or a gas is heated its temperature may rise, or a change of state may occur without change in temperature. If a given amount of heat is supplied to two different solids or to different states of the same substance then different changes in temperature may occur.

Specific heat capacity

Unit: $J \, kg^{-1} \, K^{-1}$, scalar.
Specific heat capacity c is the heat required in joules to raise the temperature of one kilogram of the substance by one kelvin.

Notice that a difference of one kelvin is the same as a difference of one degree Celsius.

Note: The specific heat capacity depends on the temperature at which it is measured. The heat required to change the temperature of one kilogram of a solid from 10°C to 20°C may not be the same as that required to change the temperature of the same one kilogram from 80°C to 90°C. However, for the accuracy required here, this variation may be neglected. For interest it may be noted that the specific heat capacity for water is $4.185 \, J \, kg^{-1} \, K^{-1}$ at 288 K.

Heat capacity

Unit: $J \, K^{-1}$, scalar.
The heat capacity of an *object* is the heat required to raise its temperature by one kelvin.

The heat capacity thus refers to a particular object as opposed to unit mass of a given substance.

Specific latent heat of fusion

Unit: $J \, kg^{-1}$, scalar.
The specific latent heat of fusion l is the heat required in joules to change the state from solid to liquid of one kilogram of the substance at its normal melting point.

Specific latent heat of vaporization

Unit: $J \, kg^{-1}$, scalar.
The specific latent heat of vaporization l is the heat required in joules to change the state of one kilogram of the substance from liquid to vapour at its normal boiling point.

Note: The specific latent heat varies with pressure and is therefore usually quoted for standard pressure.

In calculations the heat supplied, for example, by an electrical heater, will either yield a temperature rise ΔT and/or a change of state. Unknown quantities can be calculated using $H = mc\Delta t$ and $H = ml$, where c is the specific heat capacity and l is the specific latent heat.

Experiment 6.1 *To study change in temperature with time*

A fixed mass of substance is heated by an electrical heater which has a constant power output, that is the energy supplied per second is constant. A graph of temperature against time is plotted (Figure 6.2).

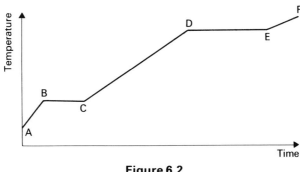

Figure 6.2

The time interval for a given temperature rise or change of state, or phase, is proportional to the heat required.

From the graph (Figure 6.2):

AB —solid state	B —melting point	BC —change of state
CD—liquid state	D—boiling point	DE—change of state
EF —vapour state		

As the time BC is less than DE, the mass remaining constant and the power supply uniform, the specific latent heat of fusion is less than the specific latent heat of vaporization. Also, as AB is steeper than CD, the specific heat capacity for the solid is less than that for the liquid, a larger rise in temperature being obtained by the solid for less heat supplied.

Changes in **volume** will occur when a mass is heated. Referring to Figure 6.2, as the temperature of the solid is increased there will be a small expansion, from A to B. At the melting point, the volume of the liquid at C will be larger than the volume of the solid at B, unless the substance is water! As the temperature of the liquid rises there will be another increase in volume, CD. The gas at E will have a much larger volume than the liquid at D.

Energy considerations are also important during this experiment. Energy in the form of heat is supplied at a constant rate. During AB there is increased vibrational energy of the solid. Between B and C the energy is required to break down the structure of the solid giving the less ordered, more mobile, liquid phase. During CD there will be an increase in kinetic energy of the moving liquid molecules. The energy absorbed at DE is required to break down the liquid structure and provide the kinetic energy for the fast moving gaseous molecules.

Example: A 20 W heater is used to supply heat to 20 g of ice and a graph similar to Figure 6.2 is obtained. What are the time intervals BC, CD, and DE, given: $l_{\text{fusion of ice}} = 3.3 \times 10^5 \, \text{J} \, \text{kg}^{-1}$, $c_{\text{water}} = 4.2 \times 10^3 \, \text{J} \, \text{kg}^{-1} \, \text{K}^{-1}$, $l_{\text{vaporization of water}} = 2.3 \times 10^6 \, \text{J} \, \text{kg}^{-1}$?

Heat supplied $= 20 \times t$ J where t is the time required
Heat required to melt the ice $= 0.02 \times 3.3 \times 10^5$ J

Assuming no heat loss

$$20 \times t = 0.02 \times 3.3 \times 10^5$$
$$t = 5\tfrac{1}{2} \text{ minutes}$$

Heat required to change the temperature from 0°C to 100°C
$$= 0.02 \times 4.2 \times 10^3 \times 100 \, \text{J}$$

Assuming no heat loss

$$20 \times t = 0.02 \times 4.2 \times 10^3 \times 100$$
$$t = 420 \, \text{s}$$
$$= 7 \text{ minutes}$$

Heat required to evaporate the water
$$= 0.02 \times 2.3 \times 10^6 \, \text{J}$$

Assuming no heat loss

$$20 \times t = 0.2 \times 2.3 \times 10^6$$
$$t = 2300 \, \text{s}$$
$$= 38 \, \text{min. } 20 \, \text{s}$$

Hence the time intervals BC, CD, and DE are $5\tfrac{1}{2}$ minutes, 7 minutes, and 38 minutes 20 seconds respectively.

Effects of heat loss

In many experiments all the heat supplied is not absorbed by the substance but some heat is lost to the container and the surroundings and therefore incorrect results will be obtained. These may be minimized either (1) by attempting to reduce heat loss or (2) by accounting for the heat loss in a control experiment.
(1) Heat loss may be reduced by avoiding draughts which lead to heat loss by convection and by using a container of a material such as polystyrene (which does not readily absorb or conduct heat) or by surrounding the container, whose heat capacity is accounted for, by an insulating jacket.
(2) For experimental temperatures below room temperature the heat absorbed from the surroundings in the time the experiment is carried out may be determined in a control experiment. For temperatures above room temperature the rate of cooling of an equal mass may be determined and hence the average heat loss estimated.

Note: Liquids should be stirred before temperatures are taken and attempts made to heat solids as uniformly as possible.

Problems

Use specific heat capacity of water $= 4.2 \times 10^3 \, \text{J} \, \text{kg}^{-1} \, \text{K}^{-1}$ and specific latent heat of vaporization of water $= 2.3 \times 10^6 \, \text{J} \, \text{kg}^{-1}$.

6.1 A container of heat capacity $120\,\text{J K}^{-1}$ contains $400\,\text{g}$ of water. What is the power of a heater which raises their temperature by $10°C$ in 1 minute?

6.2 A $20\,\text{W}$ heater was placed inside a plastic vessel containing $0.3\,\text{kg}$ of water for 2 minutes. If the initial temperature of the water was $20°C$, what was the final temperature of the water? How would your answer vary if the container were made of copper?

6.3 A heater was found to raise the temperature of $0.6\,\text{kg}$ of water from $20°C$ to $30°C$ in 1 minute. Estimate, to the nearest minute, how long it would take for this heater to evaporate $0.03\,\text{kg}$ of water initially at $80°C$.

B: Molecular Spacing, Pressure, and The Gas Laws

6:B.1 Molecular and Atomic Spacing

A particle model

The **particle model** (or *hard sphere model*) views atoms or molecules as small hard elastic spheres. For gases, this model considers the large number of fast-moving gaseous molecules and applies the laws of mechanics to obtain various relationships. Before the quantitative behaviour of gases is considered the individual physical quantities of density, mass, amount of substance, and pressure will be defined.

Cubic cells

When considering the size of a molecule, it is convenient to divide the total space available into imaginary 'cubic cells', each cell containing one molecule. In this way, it is possible to discuss the volume of one molecule and the distance between centres. Thus the cube root of the volume AB will equal the distance between centres QR, in Figure 6.3.

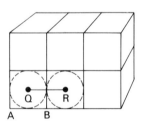

Figure 6.3   Cubic cells

Density

Unit: kg m^{-3}, scalar.

The density of a substance is defined as the mass per unit volume of that substance. Accordingly, in SI units, the density is the mass in kilograms of one cubic metre of substance.

Examples of densities (kg m^{-3})

Gases		Liquids		Solids	
steam	0.58	water	10^3	ice	0.92×10^3
oxygen	1.43	alcohol	0.79×10^3	copper	$8.9 \ \times 10^3$
hydrogen	0.09	mercury	13.6×10^3	perspex	1.19×10^3

The density depends on:

(1) the mass of the atoms or molecules, that is the relative atomic or molecular masses,
(2) the distance between the centres of the atoms or molecules. For example, water and steam have the same molecular components, H_2O, but different degrees of associativity, such that in the liquid the molecules are closer together and hence its density is greater.

Comparison of molecular spacing using densities

From the density values above, the mass of 1 m^3 of water is 10^3 kg.

$$\Rightarrow 1 \text{ kg of water has a volume of } \frac{1}{10^3} = 10^{-3} \text{ m}^3$$

$$1 \text{ kg of steam has a volume of } \frac{1}{0.58} = 1.72 \text{ m}^3$$

If 1 kg of water is evaporated then 10^{-3} m^3 of water will yield 1.72 m^3 of steam.

$$\frac{\text{volume of steam}}{\text{volume of water}} = \frac{1.72}{10^{-3}} = \frac{1720}{1}$$

Thus the volume occupied by steam molecules is 1720 greater than that for the water molecules and the mean distance between steam molecules is $\sqrt[3]{1720} \simeq 12$ times greater than that between water molecules.

If the pressure on the steam is increased, the density will increase as the distance between the molecules decreases.

In general, the separation of molecules or atoms in the gaseous state is of the order of *ten* times that of the liquid or solid state.

The volume of a given mass of gas is of the order of 10^3 greater than the volume of the same mass in the liquid or solid state.

Experiment 6.2 *Relative volumes of solids, liquids and gases*

The volume of a certain mass of carbon dioxide is measured in the solid state, then in the gaseous state at atmospheric pressure. The change in volume when a given mass of water is converted into steam at atmospheric pressure is observed. These experiments illustrate the much larger volume occupied by a gas compared to the volume of the same mass of a solid or a liquid.

Experiment 6.3   *The density of air*

Well fitting
stopper

Strong clip

Rubber
tubing

Figure 6.4

The flask is placed on an accurate balance with the clip open and its mass recorded m_1. The rubber tubing is then connected to a vacuum pump and the air removed. The clip is closed and the mass of the empty flask determined m_2. Then air is allowed to return into the flask, the stopper removed, the flask filled with water and the clip closed. The mass of the flask plus water is recorded m_3. The air temperature is noted.

The water reading enables the volume of the empty flask to be calculated using the density of water at that temperature. Thus the density of air is determined.

$$\text{volume of flask } V = \frac{m_3 - m_2}{\text{density of water}} \qquad\qquad \text{density of air} = \frac{m_1 - m_2}{V}$$

Note: The density of air varies with temperature and pressure. The density varies *inversely* as the temperature in kelvin.

Relative atomic mass scale

Since the mass of one atom or molecule is a very small quantity when expressed in kilograms, it is convenient to set up another scale of masses, the relative atomic mass scale.

Unified atomic mass unit (u)

The mass of one atom of the isotope $^{12}_{6}C$ is exactly 12 unified atomic mass units.

Thus on the relative atomic mass scale $^{12}_{6}C$ has a value of exactly 12 and the masses of all other atoms are determined *relative* to this standard, and will not necessarily be whole numbers.

Relative atomic mass of an atom

This is defined as

$$\frac{\text{mass of the atom}}{\text{mass of } ^{12}_{6}C} \times 12$$

and has *no* units.

For **molecules** the relative atomic masses of all the constituent atoms are added together to give the relative formula mass. On this scale, the relative masses for water and methane (CH_4) are 18 and 16 respectively.

A list of relative atomic masses can be found in data books.

Mole

This is a basic SI unit for the **amount of substance**, that is the number of entities involved and should not be confused with mass.

The **mole** is *defined* as the amount of substance which contains as many elementary entities as there are carbon atoms in 12×10^{-3} kg (0.012 kg) of carbon-12. The entities may be ions, atoms, or molecules.

Avogadro's Constant, N_A

Unit: mol^{-1}, scalar.

Avogadro's Constant is the number of entities in one mole. That is, the number of atoms in 12×10^{-3} kg of carbon-12.

$$N_A = 6.022\ 169 \times 10^{23}\ mol^{-1}$$

Note: This implies that the relative atomic mass or formula mass expressed in 10^{-3} kg is a mole of that substance and contains 6.022×10^{23} particles.

Examples: The relative atomic mass of tin is 119, thus 119×10^{-3} kg of tin contains 6.022×10^{23} atoms of tin.

For water 18×10^{-3} kg of water will contain 6.022×10^{23} molecules of water.

Solids—Atomic spacing using Avogadro's constant

For any species, the relative atomic or molecular mass expressed in grams, or 10^{-3} kg, is one mole.

Example: Determine the distance between centres of the copper atoms in the solid state, using $N_A = 6 \times 10^{23}\ mol^{-1}$.

Copper has a density of 9×10^3 kg m^{-3} and a relative atomic mass of 64.

First, determine the mass of one atom keeping all masses in kg.

$$6 \times 10^{23} \text{ atoms have a mass of } 64 \times 10^{-3}\ kg$$
$$1 \quad \text{atom has a mass of } 1.07 \times 10^{-25}\ kg$$

Next, use the density to determine the volume of one atom;

$$\text{density} = \frac{\text{mass}}{\text{volume}} \text{ hence the volume of one atom} = \frac{1.07 \times 10^{-25}}{9 \times 10^3} = 1.19 \times 10^{-29}\ m^3$$

Lastly, using the assumption that atoms occupy cubic cells and the length of a cell equals $\sqrt[3]{\text{volume}}$

$$\text{length of cubic cell} = \sqrt[3]{11.9 \times 10^{-30}}\ m$$
$$= 2.3 \times 10^{-10}\ m$$

Thus the distance between centres of the copper atoms = 0.23 nm.

This separation may be determined experimentally using X-ray diffraction, the results of which are found to be of the order predicted above.

Gases — Molecular spacing and molecular movement

Avogadro's Law (see Section 6:C.1) states that equal volumes of ideal gases under the same conditions of temperature and pressure contain the same number of particles. Hence, one mole of any gas containing 6.022×10^{23} particles will occupy the same volume. This volume, termed the **molar volume**, is found to be $22.4 \times 10^{-3} \, m^3$ at s.t.p.* and varies with temperature and pressure.

$$6 \times 10^{23} \text{ particles of any gas occupy } 22.4 \times 10^{-3} \, m^3$$

$$1 \text{ particle of a gas occupies } \frac{22.4 \times 10^{-3}}{6 \times 10^{23}} \, m^3$$

$$= 3.7 \times 10^{-26} \, m^3$$

Hence, the mean separation of the particles

$$= \sqrt[3]{3.7 \times 10^{-26}}$$

$$= 3.3 \times 10^{-9} \, m$$

In this calculation the approximate value of $6 \times 10^{23} \, mol^{-1}$ is used for Avogadro's Constant.

The mean separation of particles in the gaseous state is thus of the order of $3 \times 10^{-9} \, m$ or $3 \, nm$. That is *ten* times the separation of that in the solid state, which is in agreement with the density predictions.

This leads to the question, do the actual molecules 'expand' by this factor of ten, or do the comparatively small molecules move around in largely empty space? Experimental evidence indicates the latter view; gases are *easily* compressible and the gas molecules exert a pressure by *bombardment* on the walls of a container. This suggests that gas molecules are small particles moving around in 'empty' space with about nine to ten molecular diameters between neighbouring molecules.

The continuous random motion of gaseous molecules is illustrated by **Brownian movement**.

Experiment 6.4   *Brownian movement*

Smoke particles are illuminated by light and observed through a microscope. The irregular motion of one smoke particle is due to *bombardment* by the fast-moving air molecules. Since the path of the smoke particles is random it can be inferred that air molecules are in *continuous random* motion.

The phenomenon of **diffusion** also provides evidence that gaseous molecules are in motion. If two cylinders containing different gases, which do not react chemically, are joined by a connecting tap, it is found that the gases will *diffuse* into each other's container. The *rate* of diffusion will depend on the relative speeds of the molecules, the lighter molecules having the faster speeds, (see Section 6:C.1).

---

* As the volume of a gas depends on its temperature and pressure, a standard temperature and pressure, s.t.p., is defined as $273.15 \, K$ and $101 \, 325 \, Pa$, the latter being equivalent to the pressure exerted by a $0.76 \, m$ column of mercury.

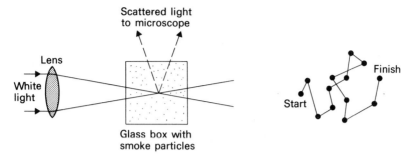

(a) Experimental arrangement　　　　(b) Path of one smoke particle

Figure 6.5 Brownian movement

Liquids—Molecular spacing

The densities of liquids are of the same order as those of solids, which indicates that the separation of particles in the liquid state is also of the order of 3×10^{-10} m or 0.3 nm. In the case of liquid water, H_2O, the distance between oxygen atoms is observed to be 0.23 nm.

Experiment 6.5 *Oil-film experiment; Langmuir trough*

The bottom of a shallow trough is covered with clean water, and lycopodium powder is sprinkled on the top. The diameter of a small drop of oil on the end of a wire loop is measured just before it is released gently on to the surface of the water. By measuring the area of spreading the thickness of the molecular layer may be determined.

$$\text{Volume of drop} = \text{area} \times \text{thickness}$$
$$\text{Diameter of drop} = d$$
$$\therefore \quad \text{Volume of one drop} = \frac{4\pi\left(\dfrac{d}{2}\right)^3}{3}$$
$$\text{Area of spreading} = A$$
$$\text{Thickness of oil} = \frac{4\pi\left(\dfrac{d}{2}\right)^3}{3 \times A}$$

If it may be assumed that a unimolecular layer is obtained and that the molecules are approximately spherical, the thickness gives the diameter of one molecule or the distance between centres.

However, many organic molecules are *not* spherical but are composed of a 'chain' of atoms such that the length of a molecule may be very much greater than its width. Many of these molecules tend to orientate themselves vertically in the surface of the water. Hence the thickness of the layer in the above experiment will give the *length* of the 'chain' of the molecule. For example, oleic acid, $C_{17}H_{33}COOH$, which is often used in Experiment 6.5, might give values of over 10^{-9} m, that is one order greater than that expected for the liquid state of inorganic molecules.

6:B.2 *Pressure*

Unit: pascal (Pa), scalar.
Pressure is the force acting on unit area. It describes the distribution of a force over a surface and hence is a scalar quantity.

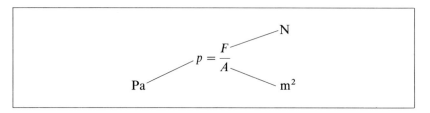

Example: The pressure on the ground due to a 72 kg man standing on stilts, each of area of cross section 30 cm^2 will be $\dfrac{72 \times 10}{2 \times 30 \times 10^{-4}}$ which is 1.2×10^5 Pa. Compare this with his pressure when standing on shoes each of area of cross section 200 cm^2. Now his pressure is $\dfrac{72 \times 10}{2 \times 200 \times 10^{-4}}$ which is 1.8×10^4 Pa, almost ten times less, as the area over which the force is distributed is greater when the man is wearing shoes.

Pressure in liquids

The pressure in a liquid depends on the density and is a function of the depth beneath the surface of the liquid. At a depth h in a liquid of density ρ the force on an area A *in* the liquid is equal to the weight of liquid on A.

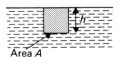

Figure 6.6 Pressure at depth h

Weight of liquid on A = volume × density × acceleration due to gravity

$$= Ah\rho g$$

$$\text{Pressure} = \frac{\text{force}}{\text{area}} = \frac{Ah\rho g}{A}$$

$$p = h\rho g$$

Pressure at a depth h in a liquid of density ρ equals $h\rho g$.

Note: The above derivation gives the pressure at a depth h due to the liquid only. The total pressure at a depth h will equal $h\rho g$ + the pressure on the *surface* of the liquid.

Since $p = h\rho g$, pressure could be stated in terms of the height of a column of a particular liquid, assuming g is constant. Mercury is a liquid used in some pressure measuring instruments because of its *high density*. Thus pressures are often quoted in mm or m of mercury, (see Section 8:C.1).

Example: Calculate the pressure in pascals equivalent to standard atmospheric pressure of 0.76 m of mercury. Take $g = 9.8 \, \mathrm{m \, s^{-2}}$ and ρ mercury $= 13.6 \times 10^3 \, \mathrm{kg \, m^{-3}}$.

$$\begin{aligned} \text{pressure} &= h\rho g \\ &= 0.76 \times 13.6 \times 10^3 \times 9.8 \\ &= 1.01 \times 10^5 \, \mathrm{Pa} \end{aligned}$$

The *approximate* value of standard pressure is thus 10^5 Pa and this may be used in problems. The exact value is 1.01325×10^5 Pa.

6:B.3   *Instruments for Measuring Pressure*

Manometer

An open tube is partly filled with mercury and is connected to the gas under investigation. The pressure of the gas at A will be the same as the pressure at B since they are at the same level.

$$\begin{aligned} \text{Pressure of gas} &= \text{pressure due to column } h \text{ of mercury} \\ &\quad + \text{atmospheric pressure, A.P.} \\ &= h\rho g + \text{A.P.} \end{aligned}$$

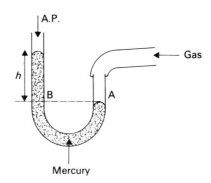

Figure 6.7   A manometer

If the pressure of the gas is less than atmospheric then the level B will be below that of A. If this difference in levels is h' then

$$\text{Pressure of the gas} = \text{A.P.} - h'\rho g$$

A water or oil-filled manometer will give larger values of h as the density of water or oil is less than that of mercury, hence small pressure *differences* may be measured.

Bourdon gauge

When a pressure is applied the metal tube tends to uncoil. Since one end of the tube A is fixed, this uncoiling causes the free end B to move, (Figure 6.8). This movement is transmitted to a pivoted quadrant gear which engages with a pinion carrying the pointer. Hence the pointer moves across a scale. This instrument thus measures pressures directly, in terms of the force on a given area, and not indirectly, as the height of a liquid column, as in the manometer or barometer.

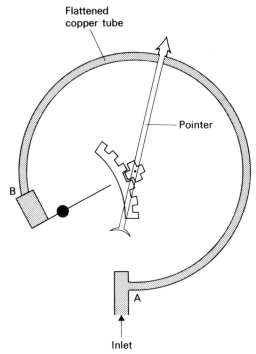

Figure 6.8   A Bourdon gauge

Experiment 6.6   *Calibration of the Bourdon gauge*

The position of the pointer on the Bourdon gauge with the exit E open to the atmosphere is noted, (Figure 6.9). This will equal the atmospheric pressure which may be read from a Fortin barometer. This reading, in mm Hg, requires conversion into Pa. The piston of a glass syringe is attached to a spring balance graduated in newtons and the other end of the syringe is joined by a piece of rubber tubing to the Bourdon gauge at E. The spring balance is extended, its reading F being the force applied to the piston of the syringe, which will move outwards reducing the pressure in the gauge.

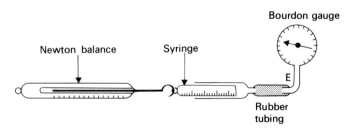

Figure 6.9   *Calibration of a Bourdon gauge*

The position of the pointer is noted for each reading F in newtons on the balance. The pressure for each position being $\dfrac{F}{A}$, where A is the area of cross section of the piston. Thus the scale may be calibrated for pressures below atmospheric.

For pressures above atmospheric the piston of the syringe must be extended before connection to the gauge, and the syringe placed vertically. It is pushed in by loading it with known masses m and the position of the pointer is noted each time. Again the pressure is equal to $\dfrac{F}{A}$ where $F = mg$.

Barometer *– for measuring atmospheric pressure*

A simple barometer is made by inverting a tall tube, *full* of mercury, into a small trough of mercury. The tube must be considerably longer than 760 mm. The

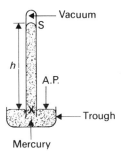

Figure 6.10 A simple barometer

mercury in the tube falls to a height of about 760 mm, leaving a vacuum at the top. Atmospheric pressure on the outside surface of the mercury in the dish will equal the pressure at the same level X inside the tube due to the column of mercury.

$$\begin{aligned}
\text{A.P.} &= \text{pressure at X} \\
&= h\rho g \text{ Pa} \\
&= h \text{ mm of mercury (mm Hg)}
\end{aligned}$$

Fortin barometer

In the simple barometer a scale cannot be placed at the upper mercury level S to record the pressure each day because when the mercury level rises inside the tube the level in the dish will fall, hence the reading at S will not be the total head of mercury as the lower level will not now be at zero. The Fortin barometer is a more accurate, direct reading, instrument, (Figure 6.11).

The 'zero' is set by adjusting the bottom screw so that the pointer P just touches the lower mercury surface, see Figure 6.11(b). The atmospheric pressure can then be read off a *permanent* vernier scale adjusted on to the upper mercury meniscus.

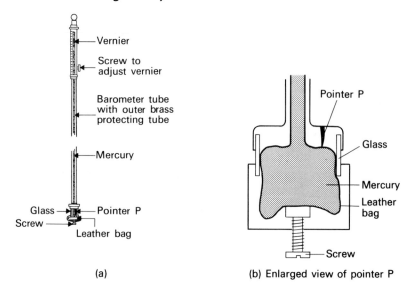

Figure 6.11 Fortin barometer

Aneroid barometer

This consists of a partially evacuated container with a flexible top. As the pressure varies the top will move in and out slightly, this movement being magnified by a series of levers to turn a pointer across the scale. This barometer has the advantage over the Fortin barometer or manometer of being more robust and portable, hence its use as an altimeter.

6:B.4 *Gas Laws*

These are the relationships between the pressure p, the volume V, and the temperature T for ideal gases.

An **ideal gas** is one in which the interactions between the gas particles are considered negligible. The following empirical laws have been discovered and constitute the behaviour of an ideal gas. Thus an ideal gas is one which *obeys* the general gas equation, namely that $\dfrac{pV}{T}$ is a constant for a fixed mass of gas.

Boyle's Law

For a fixed mass of gas at a constant temperature, the pressure p varies inversely as the volume V.

$$p \propto \frac{1}{V}$$

(constant mass and temperature)

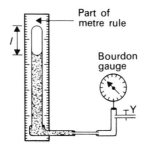

Figure 6.12 To verify Boyle's Law

Experiment 6.7 *Verification of Boyle's Law*

A certain mass of gas is trapped at the top of the tube and the pressure of the gas measured directly upon a Bourdon gauge (Figure 6.12). The volume of the gas may be obtained knowing the length l and the area of cross section of the tube A, which may be assumed constant over the length.

$$V = l \times A \qquad V \propto l \quad \text{and} \quad \frac{1}{V} \propto \frac{1}{l}$$

The pressure of the gas is varied by means of a pump connected at Y. Readings of l are taken for different pressure readings p and a graph of $\frac{1}{l}$ and p is plotted. A straight-line graph implies $p \propto \frac{1}{l}$, hence $p \propto \frac{1}{V}$, and Boyle's Law is verified if the temperature has remained constant, see Figure 6.15(a).

Charles' Law

For a fixed mass of gas at a constant pressure, the volume V varies directly as the absolute temperature T.

$$V \propto T$$

(constant mass and pressure)

Experiment 6.8 *Verification of Charles' Law*

The volume of the fixed mass of gas in Figure 6.13 is given by the length times the area of cross section of the tube, assuming negligible curvature. The pressure is equal to atmospheric pressure plus the pressure due to the thread of mercury. Providing the tube remains vertical the total pressure will not change.

The temperature of the water bath is varied and time allowed for the gas to attain this temperature before readings of volume and temperature are recorded. This is repeated for a range of temperatures. A graph of volume V, or length of tube occupied by the gas, against absolute temperature T, is plotted. A straight line through the origin implies $V \propto T$ thereby verifying Charles' Law, see Figure 6.15(b).

Note: The temperature T is on the absolute temperature scale in units of K.

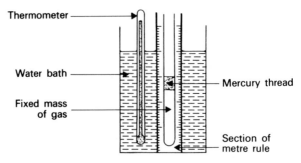

Figure 6.13   To verify Charles' Law

Pressure Law

For a fixed mass of gas at a constant volume, the pressure p varies directly as the absolute temperature T.

$$p \propto T$$

(constant mass and volume)

Experiment 6.9   *Verification of the Pressure Law*

The volume V of the gas remains constant, being equal to the volume of the gas in the container plus the gas in the tubing leading to the Bourdon gauge (Figure 6.14).

The amount of gas in this connecting tubing must be kept to a minimum as this gas will not always be at the temperature T of the water bath. The pressure p of the gas for different temperatures is then recorded, ensuring that sufficient time is allowed for all the gas to attain that temperature. A graph of p against T should be a straight line through the origin implying $p \propto T$ so verifying the Pressure Law, see Figure 6.15(c).

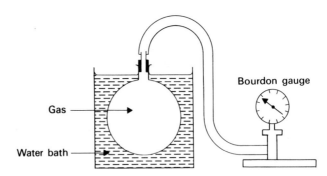

Figure 6.14   To verify the Pressure Law

Note: In Experiments 6.7–6.9 it is observed that the laws are not followed at high pressures, low volumes or high temperatures, for then the intermolecular interactions come into dominance and the gas is no longer ideal.

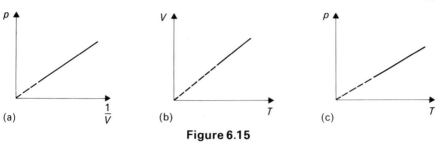

Figure 6.15

General gas equation

On combining the three gas laws a general gas equation may be obtained. Since $p \propto \dfrac{1}{V}, p \propto T$, and $V \propto T$ then

> For a fixed mass of gas $\dfrac{pV}{T}$ is a constant.

It is important to remember that the above equation $\dfrac{pV}{T} \propto$ constant, or $pV = \text{constant} \times T$, is obtained from *experimental observation*.

For an initial state $p_1 V_1 T_1$ and a final state $p_2 V_2 T_2$ clearly

$$\frac{p_1 V_1}{T_1} = \frac{p_2 V_2}{T_2}$$

This equation may be used to calculate unknown pressures, volumes or temperatures.

Example: A certain mass of gas has a volume of $50 \, \text{cm}^3$ when the temperature is $17°C$, and the barometer reading is $74 \, \text{cm}$ of mercury. What is the volume of the gas at s.t.p.?

$$V_1 = 50 \, \text{cm}^3 \qquad T_1 = 273 + 17 = 290 \, \text{K} \qquad p_1 = 74 \, \text{cm Hg}$$
$$V_2 = ? \qquad\qquad T_2 = 273 \, \text{K} \qquad\qquad\quad p_2 = 76 \, \text{cm Hg}$$

$$\frac{74 \times 50}{290} = \frac{76 \times V_2}{273}$$

and
$$V_2 = 45.8 \, \text{cm}^3$$

Note: The units must be *consistent*. They do not always need conversion into the standard units. All temperatures, however, *must* be in K, a factor of 273 being *added* to a temperature in degrees Celsius, °C.

Partial pressures

If two or more gases which do not chemically react are present in the same container the total pressure is the sum of the partial pressures which each gas would exert if alone in the container.

This statement is known as Dalton's Law of Partial Pressures.

Example: A 7 litre container of nitrogen at 10^5 Pa is joined by a connecting tap to a 4 litre container of oxygen at 3×10^5 Pa. Assuming no change in temperature, what is the resulting pressure of the mixture?

Using Boyle's Law $\qquad\qquad\qquad p_1 V_1 = p_2 V_2$

Partial pressure p_N of the nitrogen in the total 11 litres:

$$10^5 \times 7 = p_N \times 11$$

$$p_N = \frac{7 \times 10^5}{11}$$

Partial pressure p_O of the oxygen in the 11 litres:

$$3 \times 10^5 \times 4 = p_O \times 11$$

$$p_O = \frac{12 \times 10^5}{11}$$

$$\begin{aligned}
\text{Total pressure} &= p_N + p_O \\
&= \frac{7 \times 10^5}{11} + \frac{12 \times 10^5}{11} \\
&= 1.7 \times 10^5 \text{ Pa}
\end{aligned}$$

Problems

Use $g = 10\,\mathrm{m\,s^{-2}}$

6.4 Calculate the pressure due to (a) 0.8 m of mercury, (b) 5 m of water.

6.5 Determine the mass in kg of one atom of the following: tin, aluminium, platinum, and cobalt.

6.6 A Bourdon gauge reads 0.8×10^5 Pa when connected to a newton balance extended to the 8 N mark. Calculate the diameter of the piston of the syringe which joins the balance to the gauge. What would the gauge read if the syringe was removed, extended, and connected to the gauge, then placed vertically and loaded with a 400 g mass. Assume atmospheric pressure is 10^5 Pa.

6.7 A container of volume 40 cm³ is joined by a connecting tube to a Bourdon gauge which has a reading of 1.10×10^5 Pa. When the volume of the container is reduced *by* 10 cm³ the gauge reads 1.35×10^5 Pa. Calculate the volume of the connecting tube.

C: Kinetic Theory of Gases

6:C.1 Derivation of the Gas Laws using the Kinetic Theory of Gases

That a gas is observed to occupy the whole volume of a vessel rather than a part of it indicates that the gaseous molecules are in a state of motion. Furthermore, as gases are readily compressible it is reasonable to consider that the distance between the individual molecules is quite large. Brownian movement illustrates the random motion of the molecules.

A gas contains a *large* number of molecules. The movement of *individual* molecules would be difficult to calculate even if all the initial velocities and directions were known. However, since the number of molecules is *very large*, the overall, most likely, effects can be deduced using statistics. The **particle model** considers the very *large* number of *particles* moving in random directions.

The **kinetic theory** uses this particle model and, on the basis of a number of assumptions, equations are derived using the laws of mechanics.

In the kinetic theory such observations form the basis for a number of assumptions from which equations to interpret the behaviour of gases may be derived.

Assumptions of the kinetic theory

(1) The molecules behave as if they are hard, smooth, perfectly elastic point particles. (This implies no loss of kinetic energy in collisions.)
(2) The molecules are in continual random motion, colliding with the walls of a container and with each other.
(3) The volume of the actual molecules is negligible compared with the total volume they occupy.
(4) The forces of attraction between the molecules are negligible.
(5) The effects of gravity can be ignored.

(It may be shown that the mean translational kinetic energy of the molecules is proportional to the absolute temperature. This is not an assumption.)

Derivation of $pV = \frac{1}{3}Nm\overline{c^2}$ *

Consider a single molecule of gas in a container of length l and area of cross section A.

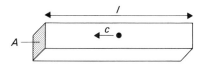

Figure 6.16   Model for the kinetic theory

If the molecule has a mass m and velocity c, parallel to the length l, its momentum as it approaches the shaded face is mc. If the collision with the walls is elastic, the molecule will leave this shaded face with momentum $-mc$.

Therefore the change in momentum is $mc - (-mc) = 2mc$.

The time between collisions on the shaded face is the time taken to travel the length of the box and back, a distance $2l$.

$$\text{Time taken} = \frac{\text{distance}}{\text{velocity}}$$

$$= \frac{2l}{c}$$

* The proof is not required for examination purposes.

From Newton's Second Law,

$$\text{Force} = \text{rate of change of momentum}$$

$$= \frac{\text{change of momentum}}{\text{time}}$$

$$= \frac{2mc}{2l/c}$$

$$= \frac{mc^2}{l}$$

Pressure is force per unit area.

$$\text{Pressure} = \frac{\text{force}}{\text{area}}$$

$$= \frac{mc^2}{lA}$$

$$= \frac{mc^2}{V}$$

where V is the volume of the container.

It is now assumed that the box contains N molecules. On average, the same number of molecules will be moving along the box as up and down or to and fro across the box. Hence the pressure on any wall is due only to $\frac{1}{3}N$ of the molecules present.

$$\text{Pressure } p = \frac{1}{3}\frac{Nmc^2}{V}$$

All the molecules do not have the same velocities and the term c^2 is usually written as $\overline{c^2}$ to show that it is an average of all the c^2 terms. It is called the **mean square velocity** of the molecules.

Note: The mean square velocity is the average of the squares of the velocities, and *not* the square of the average which would be written $\bar{c}^2$.

$$\text{Pressure } p = \frac{1}{3}\frac{Nm\overline{c^2}}{V}$$

or

$$pV = \frac{1}{3}Nm\overline{c^2}$$

This equation can be rearranged as follows:

$$pV = \frac{1}{3}Nm\overline{c^2} = \frac{2}{3}N(\frac{1}{2}m\overline{c^2})$$

giving $pV \propto \frac{1}{2}m\overline{c^2}$ for a fixed mass of gas containing N molecules.

The experimental law is $pV \propto T$ for a fixed mass of gas, where the temperature T is in kelvin.

Compare the equation $pV \propto \frac{1}{2}m\overline{c^2}$ *derived* from the assumptions of the kinetic theory, and $pV \propto T$ the *experimentally* verified law. For these two equations to be consistent, it must be concluded that $T \propto \frac{1}{2}m\overline{c^2}$, that is the **mean translational kinetic energy** $\bar{E}_k$ is directly proportional to the kelvin temperature.

$$T \propto \bar{E}_k \quad \text{and} \quad T \propto \frac{1}{2}m\overline{c^2} \quad \text{since} \quad \bar{E}_k = \frac{1}{2}m\overline{c^2}$$

Note:
(1) The average kinetic energy of all gases are the same at any given temperature; $\bar{E}_k \propto T$.
(2) For a given gas, the mean square velocity is directly proportional to the kelvin temperature; $\overline{c^2} \propto T$ (m is constant).
(3) At a given temperature, $\overline{c^2}$ is inversely proportional to the mass of the molecule; $\overline{c^2} \propto \dfrac{1}{m}$ or $m_A \overline{c_A^2} = m_B \overline{c_B^2}$ (T is constant).

Since it is the ratio of the masses m_A and m_B which is important, it is easier to use the relative molecular masses m_A and m_B and *not* masses in kilograms.

Example: The root mean square velocity of hydrogen at 15°C is 1875 ms^{-1}. Determine (a) $c_{r.m.s.}$ of hydrogen at 100°C. (b) $c_{r.m.s.}$ of oxygen at 15°C, (c) the ratio of $\bar{E}_k$ of oxygen at 100°C to that of hydrogen at 0°C. ($c_{r.m.s.}$ is defined on page 201.)

(a) Using note (2) above $\dfrac{\overline{c^2}(A)}{\overline{c^2}(B)} = \dfrac{T(A)}{T(B)}$ where A refers to 15°C (288 K) and B refers to 100°C (373 K)

giving $\dfrac{1875^2}{\overline{c^2}(B)} = \dfrac{288}{373}$

$\qquad c_{r.m.s.}$ at 100°C = 2134 ms^{-1}

(b) Using note (3) above $m_A \overline{c_A^2} = m_B \overline{c_B^2}$ where A refers to H$_2$ and B to O$_2$

$\qquad 2 \times 1875^2 = 32 \times \overline{c^2}$

$\qquad c_{r.m.s.}$ of oxygen at 15°C = 469 m s^{-1}

(c) $\qquad \dfrac{E_k(A)}{E_k(B)} = \dfrac{T(A)}{T(B)}$ where A refers to 100°C, and B to 0°C irrespective of the gas

$\qquad\qquad = \dfrac{373}{273}$

ratio = 1:1.37

Note: The temperature depends on $\frac{1}{2}m\overline{c^2}$ for *all* gases. So a molecule with a large mass will have a smaller value of $\overline{c^2}$ than one with a lighter mass at the same temperature. Hence a gas with a lighter mass will **diffuse** more rapidly than one with a high molecular mass. For example, hydrogen diffuses more rapidly than carbon dioxide at the same temperature.

Derivation of the gas laws

The three gas laws may also be derived from the kinetic theory by way of the equation $pV = \frac{2}{3}N\left(\frac{1}{2}mc^2\right)$.

(1) For a constant mass of gas, N is a constant. Hence if the temperature T is constant, $\frac{1}{2}mc^2$ is constant and pV is a constant or $p \propto \dfrac{1}{V}$, which is Boyle's Law.

(2) For a constant mass and a fixed pressure, N and p are constant so that $V \propto \frac{1}{2}mc^2$ and $V \propto T$, which is Charles' Law, since $T \propto \frac{1}{2}mc^2$.

(3) Again for a fixed mass of gas, N is constant, therefore if the volume is constant $p \propto T$, as given in the Pressure Law, since $T \propto \frac{1}{2}mc^2$.

Avogadro's Law

> Equal volumes of ideal gases at the same temperature and pressure contain the same number of particles.

This law may be derived from the kinetic theory. Consider two gases at pressures p_1 and p_2 and volumes V_1 and V_2. From the equation $pV = \frac{1}{3}Nmc^2$

$$p_1 V_1 = \frac{1}{3}N_1 m_1 \overline{c_1^2}$$

and

$$p_2 V_2 = \frac{1}{3}N_2 m_2 \overline{c_2^2}$$

The law is concerned with equal volumes at a given pressure so

$$p_1 = p_2 \quad \text{and} \quad V_1 = V_2$$

$$\Rightarrow \frac{1}{3}N_1 m_1 \overline{c_1^2} = \frac{1}{3}N_2 m_2 \overline{c_2^2}$$

The temperature of the two gases are equal giving

$$\frac{1}{2}m_1 \overline{c_1^2} = \frac{1}{2}m_2 \overline{c_2^2}$$

$$\Rightarrow N_1 = N_2$$

Therefore the number of molecules in equal volumes are the same.

6:C.2 *Mean Square Velocities*

These may be calculated using the equation $pV = \frac{1}{3}Nmc^2$.*

$$p = \frac{1}{3}\frac{Nm}{V}\overline{c^2} \qquad\qquad \rho = \frac{Nm}{V}$$

$$= \frac{1}{3}\rho\overline{c^2} \text{ where } \rho \text{ is the density}$$

*The derivation of the formula for the mean square velocity from $pV = \frac{1}{3}Nmc^2$ should be carefully learnt.

$$\overline{c^2} = \frac{3p}{\rho}$$

To obtain an idea of the average speed of the molecules the **root mean square velocity**, $c_{\text{r.m.s.}}$, is of interest.

$$c_{\text{r.m.s.}} = \sqrt{\overline{c^2}}$$

and

$$c_{\text{r.m.s.}} = \sqrt{\frac{3p}{\rho}}$$

Example: Determine the root mean square velocity of the oxygen molecule at s.t.p. The density of oxygen at s.t.p. is $1.43\,\text{kg}\,\text{m}^{-3}$.

$$p = 1.01 \times 10^5\,\text{Pa} \qquad \rho = 1.43\,\text{kg}\,\text{m}^{-3}$$

$$\overline{c^2} = \frac{3p}{\rho}$$

$$= \frac{3 \times 1.01 \times 10^5}{1.43}$$

$$= 2.1 \times 10^5$$

$$\sqrt{\overline{c^2}} = 4.6 \times 10^2\,\text{m}\,\text{s}^{-1} \text{ at 0°C and normal atmospheric pressure.}$$

The velocity $c_{\text{r.m.s.}}$ of hydrogen is higher than that of oxygen due to its lower density. Because of their higher velocities the hydrogen molecules are able to escape from the Earth, hence the lack of hydrogen in our planet's atmosphere.

Note: At s.t.p. molecular velocities are of the order of 10^2 to $10^3\,\text{m}\,\text{s}^{-1}$.

Experiment 6.10 *Determination of root mean square velocities*
A molecular beam emerging from an oven is collimated by a pair of slits such that a narrow beam reaches a rotating drum, upon which a film is fixed to the inner surface (Figure 6.17). When the slits are aligned to the small opening in the drum the molecules

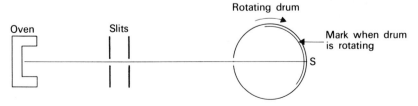

Figure 6.17 Apparatus for measuring $c_{\text{r.m.s.}}$

may enter and cross to make a mark at S, the stationary mark. If the drum is rotated the molecules will hit the film at a different point. From the distance between the marks, the diameter of the drum, and its rate of revolution, the speed of the molecules may be calculated. It is observed that the second mark made by the molecules when the drum is rotating is a broad band, dark at the centre and shading on either side, showing that the molecules do not all have the same velocity. The faster ones will reach the film nearer the mark S than the slower ones (Figure 6.18).

If the temperature of the oven is increased and the experiment repeated a mark is obtained nearer to the stationary mark S showing that the velocities of the molecules have increased. Again the deposit on the film is greatest at the centre and less on either side. However, as the temperature is further increased the *spread* of the velocities increases (Figure 6.19).

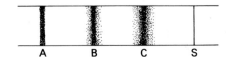

Figure 6.18   Velocity spread   **Figure 6.19** $c_{r.m.s.}$ increase with temperature

Marks A, B, and C in Figure 6.19 are for molecules leaving the oven at temperatures T_A, T_B, and T_C respectively.

Observe, that because C is nearest the stationary mark S, $T_C > T_B > T_A$.

6:C.3   *Differences between an Ideal and a Real Gas*

Gases which obey the general gas equation are termed ideal gases (see Section 6:B.4). However, at high temperatures and pressures departure from the ideal are observed. Assumptions (3) and (4) on page 197 cannot be made with such accuracy. Better agreement may be observed if account is taken of two factors.

(1) The molecules of a gas do exert a force of attraction on each other, hence the actual pressure p observed is less than that considered in the theory.
(2) The molecules themselves occupy a finite part of the overall volume of the vessel.

A more general equation to describe the properties of gases is due to Van der Waals. For one mole this equation is of the form:

$$\left(p + \frac{a}{V^2}\right)(V - b) = RT \qquad \text{where } R \text{ is a constant}$$

The term $\dfrac{a}{V^2}$ takes care of the attraction between molecules while b corrects for the effective volume occupied by the molecules themselves.

Even with this equation the behaviour observed under some conditions of temperature and pressure is not accounted for.

Note: The factors (1) and (2) above should be carefully learnt. Memorization of the equation is not important.

6:C.4   *Comparison of Solids, Liquids, and Gases*

The arrangement of the molecules in each state is the result of competition between thermal forces F_T which tend to produce collisions or vibrations and to generate disorder among molecules, and cohesive forces F_C which tend to order them. In gases, F_T is larger than F_C and therefore there is random motion. In liquids, F_T and F_C may be about the same, often giving a partly ordered structure. In solids, F_C is larger than F_T and an ordered structure usually prevails, as in the structure of crystals. Also remember that the separation of particles in the solid and liquid state is about ten times *less* than in the gaseous state. Thus the cohesive forces are acting over smaller distances in the solid and liquid states.

To change a solid into a liquid, energy must be supplied to overcome any cohesive forces. When this energy is in the form of heat it is termed the latent heat. Similarly latent heat is evolved when the liquid returns to the solid state or a gas to the liquid state. For energy considerations see Section 6:A.2.

The kinetic theory of gases does *not* apply to solids and liquids. The molecules in the solid and liquid states are not in continual random motion occupying a small volume of the total container. As mentioned above, the cohesive forces are no longer negligible. The molecules in solids do vibrate and an increase in temperature does cause an increase in vibrational kinetic energy. In liquids the molecules do have some mobility, molecules with a higher temperature will rise to the surface. Again there is an increase of kinetic energy with temperature.

Note: The kinetic theory model, a **particle model**, explains the behaviour of gases by considering the motion of the molecules. (Remember that the molecules are assumed to be small, hard, elastic spheres.) There are limitations to this model, particularly when applied to gases at high pressures, and the model does not apply to solids or liquids. Observe that the particle model provides no information about the *structure* of an individual atom or molecule.

Problems

Use density of hydrogen at s.t.p. $= 9 \times 10^{-2}\,\mathrm{kg\,m^{-3}}$, standard pressure $= 10^5\,\mathrm{Pa}$, and atomic mass of hydrogen $= 1.67 \times 10^{-27}\,\mathrm{kg}$.

Answers to two significant figures are quite sufficient.

6.8 (a) Calculate the root mean square velocity of hydrogen molecules at s.t.p.

(b) Calculate the average kinetic energy of the hydrogen molecules at s.t.p., assuming the molecular mass of hydrogen to be twice the atomic mass.

(c) What is the average kinetic energy of oxygen molecules at s.t.p.?

(d) Calculate the average kinetic energy of nitrogen molecules at 27°C.

6.9 Estimate at what temperature hydrogen molecules would have a root mean square velocity of $10^4\,\mathrm{m\,s^{-1}}$ at standard pressure.

6.10 The root mean square velocity of hydrogen at 51°C is $2.0 \times 10^3\,\mathrm{m\,s^{-1}}$. Determine the root mean square velocity at 51°C of (a) nitrogen (b) neon (c) chlorine.

7 The Atom and the Nucleus

A: The Atom

7:A.1 The Structure of the Atom

In the simplest model of the atom, the **particle** model, atoms are viewed as smooth elastic spheres. This model is useful for the kinetic theory, but it provides no details of the structure of the atom.

After Rutherford's scattering experiment (page 216) an accepted **model of the atom** consists of a central positive nucleus surrounded by a cloud of electrons. It is assumed that the electrons *move* around the nucleus. Remember, *energy* is required to move an electron away from the positive nucleus because of the electrostatic force of attraction. (The discrete nature of electric charge has been discussed, pages 50, 51.)

In Bohr's orbit model for the hydrogen atom (1913) the single electron moves around the nucleus in one of certain circular **orbits** but not in between these orbits. However, if energy is supplied, the electron can jump to an outer orbit. Also the electron can lose energy and fall to an inner orbit. Hence there are various *permitted* energy states for the atom but the energy cannot vary continuously.

Although Bohr's model does not apply to other atoms with more than one electron, the idea of energy levels *is* applicable.

Energy levels

The electrons in any atom are associated with particular energy levels. If the electrons are in the lowest possible energy levels the atom is said to be in its **ground state**. The electrons may occupy other higher energy levels, when the atom is said to be in an **excited state**.

The energy levels can be represented by a series of horizontal lines, Figure 7.1. It is important to realize that there are only certain permitted energy states. The energy is said to be **quantized** since it cannot vary continuously, but only have certain precise values.

Note: The zero of energy refers to the condition of the atom when the electron has been completely removed, hence the negative values for the energy levels.

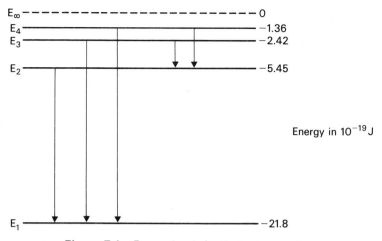

Figure 7.1 Energy levels for the hydrogen atom

Example: An analogy, to emphasize the idea of quantization of energy levels, is provided by a series of steps of variable height.

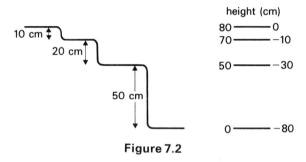

Figure 7.2

An object can only lose fixed amounts of potential energy by falling down *one* step or *several* steps. From Figure 7.2 the possible 'heights' in cm through which the object could fall are 10, 20, 30, 50, 70, and 80. Observe the negative values when zero is chosen to be at the top of the steps.

Referring back to the atomic energy levels, if energy is *supplied* the electrons can move up to higher energy states, which are represented by an arrow pointing *upwards*. Also, energy is released when a transition takes place to a lower energy level as shown by the arrows in Figure 7.1. The amount of energy absorbed or emitted depends only on the energy difference between the two levels, e.g. $E_4 - E_1$ or $E_3 - E_2$. Hence there are only a finite number of energy packets, or **quanta**, possible.

7:A.2 *Spectra*

Line spectra

When an electron passes from an excited energy level E_i to a lower energy level, or the ground state, E_j, the energy available $E_i - E_j$ is emitted as light or other electromagnetic radiation, of frequency f such that the energy difference $\Delta E = E_i - E_j$ is given by

$$\Delta E = hf$$

where h is Planck's constant $(6.63 \times 10^{-34}\,\text{J s})$.

A packet of light energy, a **photon**, is emitted, the frequency f depending only on the energy difference. The intensity of the radiation is determined by the *number* of transitions taking place, that is on the *number* of photons emitted.

Note: All the energy from a transition ΔE produces one photon of frequency f. This energy *cannot* be split up.

For any transition the frequency of the photon emitted can be calculated from the difference in the energies of the levels. For example (from Figure 7.1):

$$E_3 - E_1 = (-2.42 - (-21.8)) \times 10^{-19} = hf_1$$

giving
$$f_1 = 2.92 \times 10^{15}\,\text{Hz (in the ultra-violet)}$$

and
$$E_3 - E_2 = hf_2$$
$$f_2 = 4.57 \times 10^{14}\,\text{Hz (in the visible region)}$$

These frequencies for electronic transitions may be in the ultra-violet, visible, or infra-red regions.

Conversely an electron may be excited to a higher energy level by absorbing radiation of this frequency f, that is absorbing a photon of light energy $E_i - E_j$.

Each element has a different nucleus and a different number of electrons hence the values of the energy levels will vary between elements.

> The values of the energy levels depend on the individual atom. Therefore, the energy difference between levels and hence the frequencies which may be emitted or absorbed are **characteristic** of each atom.

Observe that each element can only emit a finite number of frequencies.

Experiment 7.1 *A line emission spectrum*

For an atom to emit a photon of light, energy must be supplied to excite that atom. Either the substance is heated, or an electrical discharge is passed through the vapour of the atoms. The emitted light is then viewed with a spectrometer (see page 156). This instrument uses a prism or a diffraction grating to split up the incident light into a spectrum. When an atomic species is used a number of coloured lines are observed. As expected the lines are different for each element. The line spectra of mercury, sodium and helium are given in Figure 7.3.

Note: A **line emission spectrum** consists of a number of coloured lines on a black background.

Because a line spectrum is characteristic of a particular atomic species it can be used to detect small quantities of those atoms, hence its use to chemists. The

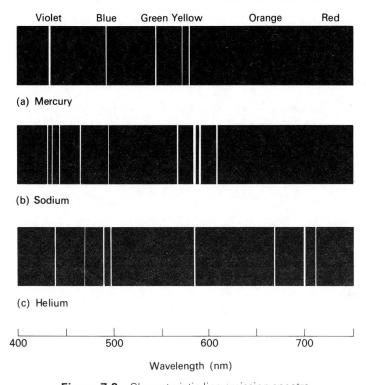

Figure 7.3 Characteristic line emission spectra

intensity of the lines are not all the same. Some transitions are more favoured than others.*

Experiment 7.2 *The line absorption spectrum of sodium*

If white light illuminates the vapour of particular atoms then those wavelengths, or frequencies, corresponding to the line spectra of the vapour atoms will be **absorbed**. Any photon of light energy of frequency f equal to the difference between two energy levels of that atom may be absorbed, resulting in an electron associated with a higher energy value.

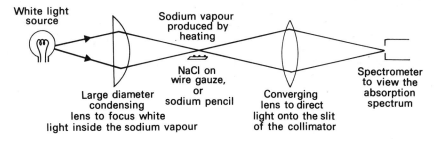

Figure 7.4 Absorption spectrum

* In atoms with more than one electron some transitions are forbidden.

Light then reaching a spectrometer will have these wavelengths *absent*, hence dark lines will be observed on a continuous spectrum background.

A dark line is observed in the yellow region. Because of the limitations of this apparatus it is not always possible to discern the two separate dark lines in the yellow region or the other dark lines in the green, blue, or violet. However, it should be remembered that all the frequencies corresponding to allowed energy differences can be absorbed.

Note: The **line absorption spectrum** has black lines on a continuous spectrum background. The frequencies corresponding to the black lines in the absorption spectrum are the *same* as the frequencies of the coloured lines in the emission spectrum.

The element helium was first detected in the *absorption* spectrum of the Sun, hence its name from the Greek *helios* meaning the Sun.

Band spectra

A **molecule** also has vibrational and rotational energy levels. Transitions between these levels results in spectra lying in the *infra-red* and *microwave* regions respectively (compare with the electronic transitions in the u.v. and visible regions). In contrast to the *line* spectra for atoms, *band* spectra are observed for molecules.

When a molecular species, a compound, is excited by being supplied with energy, a band emission spectrum is observed which is characteristic of that compound. Similarly, an absorption band spectrum can be obtained when electromagnetic radiation of the appropriate frequencies is passed through the vapour of the compound. The infra red *absorption bands* for molecules are used to identify certain groups within the molecule, as for example the carbonyl group $>C{=}O$.

Note: The frequencies of the emission band and absorption band are the same for a given compound and are characteristic of that compound.

Continuous spectrum

This is the spectrum of white light which has all, or nearly all, the wavelengths of the visible region. A gradually changing band of colour from red through to violet is observed. This may be produced by using a glowing tungsten lamp as a source and dispersing the light with a glass prism. Sunlight will give a continuous spectrum although with a few wavelengths absent.

7:A.3 Photoelectric effect

This is the emission of electrons from a substance when illuminated by electro-magnetic radiation. The particular case of electron emission from the surface of a metal when illuminated by light or ultra-violet radiation will be discussed here.

Note: Compare the terms **photoelectric emission**, which is the emission of electrons under the action of light, and **thermionic emission**, which is the emission of electrons from a heated cathode. When a valve or a solid state device contains a cathode which emits electrons under the action of light, the cathode is called a **photocathode**.

Experiment 7.3 *To demonstrate the photoelectric effect*

A clean zinc plate is attached to the plate of a negatively charged electroscope and illuminated by ultra-violet radiation, as shown in Figure 7.5. The leaf is observed to *fall* indicating that negative charge has been lost, which suggests that the zinc plate has emitted electrons. If the intensity of the ultra-violet is increased, the leaf is observed to fall much quicker.

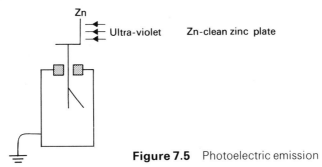

Figure 7.5 Photoelectric emission

The electroscope is recharged, with a negative charge, and the ultra-violet replaced by a strong white light source. The leaf does *not* fall.

The experiment is repeated using ultra-violet and a positively charged electroscope. Again the leaf does *not* fall.

These results show that only ultra-violet light can provide enough energy for an electron to escape from a negatively charged plate. Remember, the energy of one photon cannot be divided up, and that an electron can absorb energy from just one photon. When the intensity of the ultra-violet is increased the number of photons is increased, hence the number of electrons ejected is increased and the leaf falls rapidly.

Experiment 7.4 *Photoelectric emission*

When the zinc plate in the apparatus shown in Figure 7.6 is illuminated by ultra-violet radiation, the milliammeter shows a deflection, even when there is no p.d. between P and the zinc plate.

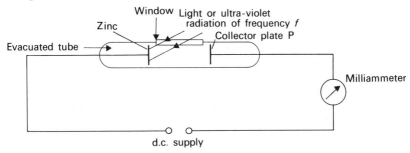

Figure 7.6

(1) Using radiation of frequency f, if the potential of plate P is increased the current increases to a maximum value I_{max} then remains constant. At the p.d. V_M for this maximum current, *all* the emitted electrons are reaching plate P. A negative potential

$-V_0$ is required to give a zero current, i.e. to prevent the most energetic electrons reaching P. Thus electrons are emitted with a *range* of kinetic energies.

An alteration of the *intensity* of the radiation does not affect the value of $-V_0$ (called the *stopping potential*) but an increased value of I_{max} will be obtained. This implies that the *number* of electrons emitted depends on the intensity, but the energy of the electrons is *not* affected.

(2) With the potential of P maintained above V_m, to ensure that any emitted electron is collected, the frequency f of the radiation is varied and the current I is recorded. A graph of I against f is plotted as shown in Figure 7.7.

For frequencies below a certain frequency f_0, termed the **threshold frequency**, no current is recorded, that is no electrons are emitted. Even if the intensity of this radiation f_0 is considerably increased the current remains at zero. However, as the frequency is increased electrons will be emitted, even at low intensities, and a current detected.

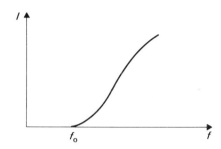

Figure 7.7   Variation of *I* with *f*

A certain amount of energy is required to eject the electron from the metal. If W_{min} is the *minimum* energy required for an electron to escape, then $W_{min} = hf_0$ and the electron will leave with zero velocity. However, many electrons just inside the surface will require *more* than this minimum to escape. The value of W_{min} depends on the metal. Brass and aluminium have larger values than zinc, hence brass and aluminium have higher threshold frequencies. When the frequency is greater than f_0 any extra energy appears as kinetic energy of the fast moving electrons. Hence the variable speeds of the electrons. The photoelectric effects are summarized below:

No electrons are emitted below a certain threshold frequency f_0 which depends on the metal irradiated.

The electrons are emitted with kinetic energies ranging from zero to a certain maximum, this maximum depending on the *frequency f* of the incident *photons*.

The number of electrons emitted depends on the *intensity* of the incident radiation, that is on the *number* of incident *photons*.

Experiment 7.4 can be used to determine the value of Planck's constant h.

Example: The minimum energy to eject an electron from a clean zinc plate is 6.1×10^{-19} J. Determine:

(a) the energies of a photon of green light ($\lambda = 550\,\text{nm}$) and a photon of ultra-violet ($\lambda = 200\,\text{nm}$) and comment on their ability to produce photoemission,

(b) the minimum frequency required to eject an electron from zinc.

(a) Using $c = \lambda f$ $f(\text{green}) = \dfrac{3 \times 10^8}{550 \times 10^{-9}}$ and $f(\text{u.v.}) = \dfrac{3 \times 10^8}{200 \times 10^{-9}}$

$$= 5.45 \times 10^{14}\,\text{Hz} \qquad\qquad = 1.5 \times 10^{15}\,\text{Hz}$$

Using $E = hf$ $E(\text{green}) = 3.6 \times 10^{-19}\,\text{J}$ $E(\text{u.v.}) = 9.9 \times 10^{-19}\,\text{J}$

The ultra-violet will eject an electron from the zinc, but the green light does not have sufficient energy.

(b) Using $E = hf$ $6.1 \times 10^{-19} = 6.63 \times 10^{-34} \times f$

giving the minimum frequency $f = 9.2 \times 10^{14}\,\text{Hz}$ (ultra-violet)

7:A.4   Wave Particle Duality

Are light and electrons waves or particles?

In Chapter 6 the wave nature of light was discussed, interference and diffraction being demonstrated. The electron has been treated as a particle obeying the laws of mechanics, but it may also show wave properties such as diffraction.

Particle aspect of electrons

(1) In a cathode ray tube the electron beam may be deflected and observed by the spot the electrons produce on a luminescent screen. The motion of the electrons is found to obey the equations of classical mechanics. (Remember that the charge on the electron is particulate in nature.)

(2) In collisions involving electrons in a cloud chamber, the changes in energy and momentum of the particles are found to obey the laws of conservation of energy and momentum in the same way as the colliding vehicles, balls, and pucks discussed in mechanics.

Particle aspect of electromagnetic radiation

The 'particles' of electromagnetic radiation, **photons**, have zero mass and energy hf where f is the frequency and h is Planck's constant. It may also be shown that they have momentum, see (3) below.

(1) Line spectra

The emission and absorption line spectra illustrate the particulate nature of light. All the energy available from a transition produces a photon of light of a given frequency and this photon cannot be divided.

(2) Photoelectric effect

The photoelectric effect can be explained only by considering the particle aspect of light. Energy cannot be absorbed continuously from a wave.

(3) Compton effect

In 1924 Compton observed that a beam of X-rays of frequency f scattered by electrons through an angle θ had a reduced frequency f'. The recoil electron, which was effectively at rest before collision, acquired energy and momentum (Figure 7.8).

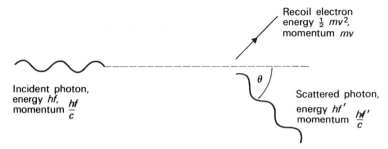

Figure 7.8 Compton effect

The conservation laws of momentum and energy were found to be obeyed if the momentum of a photon was taken as $\dfrac{hf}{c}$, where c is the velocity of light. The photons may thus be treated like particles with the collision being similar to that between two objects on a frictionless surface!

Wave aspect of electromagnetic radiation

(1) Interference of light may be demonstrated by Young's slits experiment, (see Section 5:C.2).
(2) Diffraction of light is shown by the diffraction grating (see Section 5:C.3) and by diffraction patterns obtained around sharp edges or small apertures.
(3) X-rays may be diffracted by crystals. When the crystal is in a powdered form the diffraction pattern may be a series of concentric rings, as shown in Figure 7.9.

Figure 7.9 X-ray diffraction rings produced by a crystal (*National Chemical Laboratory*)

Wave aspect of electrons

(1) *Electron diffraction*

When a beam of electrons is fired at a thin gold film a **diffraction** pattern similar to X-ray diffraction in crystals is obtained, see Figure 7.10. Compare Figures 7.9 and 7.10.

To explain the distribution of the electron pattern it is necessary to consider that the electrons have associated wave properties. It is observed that if the velocity, and therefore the momentum p of the electrons is increased the associated wavelength λ decreases, $p\uparrow$ as $\lambda\downarrow$. The relationship between the

Figure 7.10 Electron diffraction rings produced by a sheet of thin gold foil (*Science Museum, Prof. Sir G. Thomson F.R.S.*)

momentum p and the wavelength λ is found to be

$$p = \frac{h}{\lambda}$$

This relationship was originally proposed by Louis de Broglie in 1924, hence the wavelength λ is called the **de Broglie wavelength**. If $p = \frac{h}{\lambda}$ then $p = \frac{hf}{c}$ which is identical to the equation for the momentum of a photon in the Compton effect!

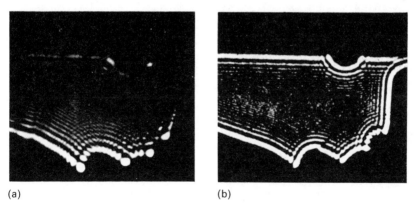

(a)   (b)

Figure 7.11 (a) Diffraction pattern produced by electrons emitted from a sharp point and passing through an aperture between tiny opaque crystals. (b) Diffraction of light passing through an opening in a metal plate cut to the same shape as the gap in the crystal in (a). (*Prof. Sir B. Pippard F.R.S., Cavendish Laboratory*)

Figure 7.11 shows the similarities between electron and light diffraction. The beam of electrons passes through an aperture between tiny crystals. The light is diffracted through a similarly shaped opening in a metal plate.

As the wavelength associated with a particle depends on its momentum diffraction effects are unlikely to be observed with particles of large mass!

(2) *Electron microscope*

An **electron microscope** is used to study minute objects such as cell nuclei. Electric fields are used for focusing the electrons but the principle and theory of

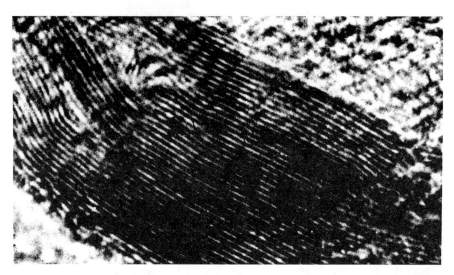

Figure 7.12 An electron microscope study of a carbon fibre. The separation between the graphite layers of 0.34 nm is readily observed. (*Dr. J. R. Fryer, Glasgow University*)

the microscope may be treated in the same manner as the light microscope. The small wavelengths associated with the electron beams give the electron microscope its ability to resolve at high magnification ~ one million (see Figure 7.12).

Summary: wave particle duality

	Light	*Electron*
Particle aspect	(1) line spectra photon emission (2) photoelectric effect photon absorption (3) Compton effect photon deflection	(1) cathode ray tube electron deflection (2) cloud chamber electron collision
Wave aspect	(1) interference (2) diffraction of light (3) X-ray diffraction	(1) electron diffraction (2) electron microscope

7:A.5 *An Atomic Model*

An atomic model with electrons at discrete energy levels is useful to interpret line spectra, but classical mechanics and electricity cannot *account* for these discrete states.

Both light and electrons can be treated as waves *and* particles. It should be mentioned that the electron cannot be directly 'seen': only its properties and effects can be observed. Some of the observations are explained by treating the electron as a particle with energy and momentum, other observations are interpreted by considering the de Broglie wavelength.

It is impossible to predict with certainty how particular electrons move, using the laws of classical mechanics. These laws apply only to particles. However, another type of mechanics, called **quantum mechanics**, can be used to calculate *observable* quantities. The electrons are associated with **orbitals**, which are mathematical wave functions. Although the position of an electron cannot be determined exactly using these orbitals, the *probability* of finding an electron within a certain volume can be calculated, using the rules of quantum mechanics.

Quantum mechanics *can* account for the discrete energy levels.

All particles have associated wavelengths but for large particles the wavelengths are *very* small and wave effects can be ignored.

Example: Determine the wavelength of an electron moving at $10^5\,\mathrm{m\,s^{-1}}$ and a 10 g bullet travelling at $100\,\mathrm{m\,s^{-1}}$.

Using the de Broglie relationship $\lambda = \dfrac{h}{\text{momentum}}$

For the electron $\lambda = 7.3 \times 10^{-9}\,\mathrm{m}$ or 7.3 nm
For the bullet $\lambda = 6.63 \times 10^{-34}\,\mathrm{m}$

Observe that the wavelength of the bullet is *very* small, too small to give any *observable* diffraction effects. Hence the wave properties of the bullet can be ignored.

In general, if the associated wavelength of a moving particle is larger than, or of the order of, the atomic size $\sim 10^{-10}\,\mathrm{m}$, then quantum mechanics must be used. Thus any satisfactory description of an atomic model requires the study of quantum mechanics.

Problems

7.1 What is the energy of a photon of the green spectral line of mercury whose wavelength is $5.46 \times 10^{-7}\,\mathrm{m}$? (Take the velocity of light as $3 \times 10^8\,\mathrm{m\,s^{-1}}$ and Planck's constant as $6.6 \times 10^{-34}\,\mathrm{J\,s}$.)

7.2 Why must the zinc plate in Experiment 8.2 be clean?

7.3 In photoelectric emission why do the kinetic energies of the emitted electrons vary, even though monochromatic light is used?

7.4 Would it be possible to demonstrate diffraction with particles or objects other than electrons?

7.5

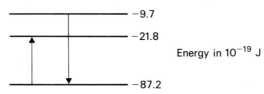

-9.7

-21.8

Energy in 10^{-19} J

-87.2

Figure 7.13 Energy levels for the helium ion

Two transitions are shown in Figure 7.13.

(*a*) Calculate the frequency of the absorption line.

(*b*) Determine the frequency of the emission line.

(*c*) State the energy of the remaining possible spectral line.

7.6 The minimum energy to produce photoemission for a certain metal plate is 8×10^{-19} J.

(*a*) Determine the minimum frequency of the radiation needed to eject an electron and state the type of radiation.

(*b*) Why is this the minimum frequency?

(*c*) How would the velocity of the ejected electron alter if the intensity of the radiation was doubled?

7.7 Describe and compare the emission and absorption spectrum of an element.

B: A Nuclear Model and Radioactivity

7:B.1 *Nuclear Structure*

Thomson's model

In this model of the atom, discussed by Thomson in 1904, the electrons were thought to be embedded in a sphere of diffuse positive material, this material filling the *whole* atomic volume. At this time some properties of the electron were known but the proton had not yet been observed.

Rutherford's model

As Rutherford was dubious of Thomson's model, Geiger and Marsden conducted the following 'scattering' experiment in 1911.

The experiment was performed in a vacuum to prevent other collisions by the α particles. The α particles, which were known to be small, were directed at a very thin gold foil target, as shown in Figure 7.14. Most α particles passed to the

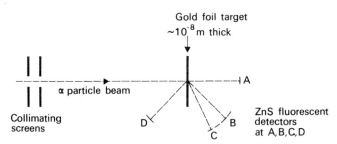

Figure 7.14 Rutherford's scattering experiment

detector at position A but some are deviated through various angles and detected at positions B to C. Occasionally (about once in 8000 cases) an α particle was repelled back and detected around position D. If the mass and charge of each gold atom were distributed throughout its volume *none* of the α particles should have been deflected backwards. This brought about a proposal by Rutherford for a new model of the atom (Figure 7.15).

In this model the positive charge and most of the mass of each atom is

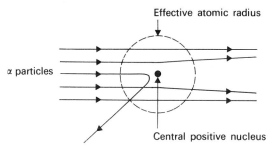

Figure 7.15 Rutherford's atomic model

concentrated in a small central nucleus. The electron cloud is around this nucleus and accounts for the volume of the atom. This dense central nucleus could then produce the large deflections when the occasional α particle passed close by.

From similar experiments values were obtained for effective nuclear and atomic radii.

Basic nuclear terms

An accepted model of the atom consists of a central nucleus with a nuclear radius of the order of 10^{-15} m surrounded by a cloud of electrons, giving an atomic radius of the order of 10^{-10} m. The nucleus contains **protons** and **neutrons**. Each proton has a positive charge equal in *magnitude* to the charge of an electron. The neutron is electrically neutral.

Nucleon is the collective name for protons and neutrons.

The **atomic number Z** of an element is the number of protons in the nucleus, which is also equal to the number of electrons in the outer cloud.

The **mass number A** is the total number of nucleons in the nucleus.

A **nuclide** is a particular nucleus with a specified number of protons and neutrons. Any nuclide may be represented by its chemical symbol X together with its atomic and mass numbers; $_{Z}^{A}\text{X}$, for example: $_{1}^{2}\text{H}$, $_{6}^{12}\text{C}$, $_{6}^{14}\text{C}$, $_{7}^{14}\text{N}$, $_{92}^{235}\text{U}$.

Isotopes are two or more nuclides of the same element, that is with the same chemical symbol X and atomic number Z, but with different numbers of neutrons hence different mass numbers A. For example $_{6}^{12}\text{C}$ and $_{6}^{14}\text{C}$ both have six protons but six and eight neutrons respectively. Hydrogen $_{1}^{1}\text{H}$ has two isotopes: **deuterium** $_{1}^{2}\text{H}$ which has one neutron, and **tritium** $_{1}^{3}\text{H}$ which has two neutrons.

Isobars are two or more nuclides of different elements with the same mass numbers, for example $_{6}^{14}\text{C}$ and $_{7}^{14}\text{N}$.

7:B.2 Radioactivity

In 1896 Becquerel discovered that certain emanations from salts of uranium affected a photographic plate. Later Rutherford in 1899 distinguished between two types of emanation, one of low penetrating power, which he called α-rays, and the other of high penetrating power, which he called β-rays. One year later Villard noticed a third type, even more penetrating, which he termed γ-rays.

Absorption of radiation

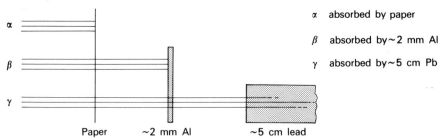

α	absorbed by paper
β	absorbed by ~2 mm Al
γ	absorbed by ~5 cm Pb

Figure 7.16   Absorption of α and β particles and γ-rays

Deflection in a magnetic field

For the same magnetic field applied perpendicular to the direction of motion of the rays, γ-rays are observed to be unaffected, α-rays are deflected slightly in a direction indicating that they are positively charged, while β-rays are strongly deflected in the opposite direction suggesting a negative charge and a lighter mass (Figure 7.17). Measurements show that the β-rays are fast moving electrons and the charge on the α-rays is double that on the β-rays. Thus α- and β-rays are streams of charged particles.

Experiment 7.5   *Rutherford and Royds' experiment to show that α particles are helium nuclei*

Radon gas, which is observed to emit α particles, is introduced into the thin-walled glass tube. The α particles escape into the evacuated thick-walled glass envelope. After several

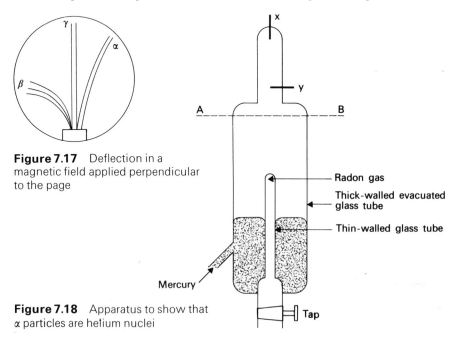

Figure 7.17   Deflection in a magnetic field applied perpendicular to the page

Figure 7.18   Apparatus to show that α particles are helium nuclei

Radon gas

Thick-walled evacuated glass tube

Thin-walled glass tube

Mercury

Tap

days the mercury is forced up to the level AB compressing the α particles, which have now captured stray electrons to become electrically neutral. A high p.d. is applied between X and Y to obtain a discharge in the gas. The spectrum emitted, which is viewed through a spectrometer, is observed to be identical to the spectrum of helium indicating that α particles are helium nuclei. An α particle therefore consists of two neutrons and two protons.

Origin of α, β, and γ radiation

α and β particles and γ-rays *originate* from the *nucleus*.

When an **α particle** is emitted, the nucleus loses two neutrons and two protons.

The **β particles** are electrons emitted from the nucleus when a neutron *changes* into a proton.

$$n \rightarrow p^+ + e^-$$

Note: An electron from the outer electron cloud is *never* emitted as a β particle.

γ-rays are high-energy electromagnetic radiation emitted when a 'rearrangement' of the neutrons and protons occurs and the nucleus 'falls' from a high energy state to a lower energy state.

This difference in energy is equal to hf, where h is Planck's constant and f is the frequency of the γ-rays.

Radioactivity is the spontaneous disintegration of the nucleus of an atom from which α, β and γ radiation may be emitted with occasionally other particles or nuclear fragments. It is caused by an unstable nucleus tending to a stable condition and is *not* affected by chemical reactions or the outer electrons. Natural sources of radioactivity include cosmic rays and radioactive rocks.

Properties of α and β particles and γ-rays

Property	α particle	β particle	γ-rays
Nature	Helium nucleus $2n + 2p$	Electron	High-energy electro-magnetic radiation
Charge	$+2$	-1	Zero
Penetrating power	Low	Medium	High
Absorption	Paper or 10^{-2} mm Al	~ 2 mm Al	~ 5 cm lead
Ionization	High	Medium	Low
Velocity	Of the order of 10^7 m s^{-1}	Up to 10^8 m s^{-1}, but variable	At velocity of light 3×10^8 m s^{-1}
Spread of velocities	One or few definite velocities	Wide spread	All the same
Example of suitable source	$^{241}_{95}$Americium	$^{90}_{38}$Strontium	$^{60}_{27}$Cobalt

For laboratory experiments a radioactive source emitting only one type of radiation is often required. The isotopes listed above are the most useful ones for each type, although $^{241}_{95}$Am does emit some γ-rays and $^{60}_{27}$Co gives a small β emission which may be removed by absorption.

When these radiations travel through a gas they cause ionization along their path. Due to the heavier mass and lower speeds the α particles produce considerable ionization per mm of their path length and quickly lose their energy, hence their tracks are short and they are easily absorbed. The β particles, with their higher velocities, cause less ionization per mm, and therefore produce longer tracks and penetrate further. The γ-rays produce very little ionization per mm, do not leave tracks and have high penetrating powers.

7:B.3   *Detection of Radioactivity*

Background radiation

Cosmic rays and radioactive materials provide a general background of radiation, which can be detected by a suitable device. It is necessary therefore to take a count or reading before and after an experiment involving a radioactive source. The average of these readings is called the **background count**. This count must be subtracted from all experimental readings to obtain results due to the activity of the source alone.

Methods for detection of radioactivity

(1) *Ionization chamber*

In the ionization chamber shown in Figure 7.19, radiation enters C, usually through a wire-mesh panel in the lid, causing ionization and therefore a small transient current between C and P. This may be detected on a milliammeter, after

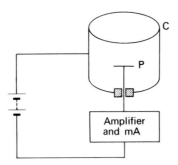

Figure 7.19   Ionization chamber

amplification. If the radioactive source is gaseous it may be introduced directly into the chamber now fitted with a full lid with no mesh panel, and its activity observed. This instrument is most suitable for the detection of α, β, and other charged particles which produce sufficient ionization along their path.

(2) *Geiger-Müller tube*

This is a sensitive ionization chamber which may be used to detect α and β particles and γ-rays.

α or β particles enter the chamber through the thin mica end-window but γ-rays may penetrate the walls of the tube (Figure 7.20(a)). The radioactive particles ionize some of the neon atoms, which accelerated by the electric field undergo further collisions. These collisions produce more ionization in sufficient quantity to give a pulse on a meter connected to the output. To prevent sparking and to clear the tube for the next event, a quenching agent, such as a trace of bromine gas, is used.

Note: One pulse is recorded for each radioactive event regardless of the energy of the ionizing particle.

Before commencing an investigation a graph of count rate against operating p.d. is plotted for the Geiger–Müller tube (Figure 7.20(b)). Below a certain p.d.,

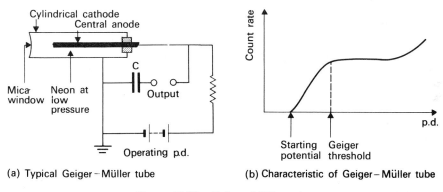

(a) Typical Geiger – Müller tube

(b) Characteristic of Geiger – Müller tube

Figure 7.20   Geiger–Müller tube

called the **starting potential**, no events are recorded. The central plateau region is stable for fluctuations in supply, hence a p.d. is chosen in this middle 'plateau' section ~ 400–$450\,$V in the type usually used in schools and colleges.

To count the total number of pulses a **scaler** may be used. A **rate meter** records the average value of the ionization current over a specified time interval, that is the average number of pulses per second, the **count rate**.

(3) *Cloud chamber*

α, β, or other charged particles ionize the gas along their path. If this occurs in a liquid about to boil or in a vapour about to condense the ions will preferentially encourage the formation of droplets or bubbles giving a track. The path of the particle is then observed and can be photographed.

(*a*) **Wilson cloud chamber:**   Water vapour in a closed chamber is expanded suddenly by a piston and thereby cools. This causes condensation on ions produced by any radioactive particle present. The expansion must be timed to occur when a charged particle enters the chamber. A photographic device can be linked to the piston and a photograph taken immediately after the expansion (see Figure 1.39 on page 40). An electric field is maintained across the chamber to clear the ions before the next expansion.

(b) **Bubble chamber:** Liquid hydrogen at its boiling point fills a chamber and charged particles are revealed by a trail of bubbles. Lighting from beneath allows the trail to be photographed.

(4) *Photographic emulsion*

α and β particles and γ-rays will blacken photographic plates and hence leave lines on photographic emulsion plates. A number of plates may be placed on top of each other to obtain a three-dimensional track.

(5) *Solid-state detectors*

The junction of a solid-state device is exposed to the radiation. The electrons and/or holes produced by the radiation will give rise to short currents which may be amplified and fed to a scaler or rate meter. By suitable choice of solid-state materials, α and β particles or γ-rays may be detected.

(6) *Scintillation counters*

Radiations produce a flash of light, a **scintillation**, when impinging on certain materials called *phosphors*, for example α particles cause zinc sulphide to scintillate. These weak flashes may be amplified by a photomultiplier. In modern scintillation counters liquid phosphors are used. By a suitable choice of phosphor most radiations may be detected by this method. This is one of the few devices which may be used to measure the *energy* of the incident radiation or particle.

7:B.4 *Disintegration*

Radioactivity is a *random* process. In a given time a certain number of parent nuclei will disintegrate into their daughter products but it is impossible to state when one particular nucleus will disintegrate.

   The **activity** of a substance is the number of disintegrations per second and is measured in **becquerels**.

$$1 \text{ becquerel (Bq)} = 1 \text{ disintegration per second } (s^{-1})$$
(The former unit, the **curie** equals 3.7×10^{10} Bq.)

Since the activity of a given sample of a nuclide is continuously varying as the nuclides decay, this concept is not always very useful.

Half life

Unit: seconds (to years), scalar.

> The half life of a radioactive nuclide is the *time taken* for half the number of parent nuclei to disintegrate; or the time taken for the activity to fall to half of its initial value.

   Although the decay is a random process the half life will be the same for all samples of the nuclide, providing that a very large number of atoms are used.

Neither chemical combination of the nuclide nor changes in its physical surroundings will alter the half life.

Half lives range from over millions of years to less than microseconds.

Decay series

Many parent nuclei decay into daughter products which are also radioactive.

Examples:

$$^{238}_{92}\text{U} \rightarrow {}^{234}_{90}\text{Th} + {}^{4}_{2}\text{He} \qquad \alpha \text{ emission}$$
$$^{234}_{90}\text{Th} \rightarrow {}^{234}_{91}\text{Pa} + {}_{-1}^{0}\text{e} \qquad \beta \text{ emission}$$
$$^{234}_{91}\text{Pa} \rightarrow {}^{234}_{92}\text{U} + {}_{-1}^{0}\text{e} \qquad \beta \text{ emission}$$

$^{234}_{92}\text{U}$ then decays by repeated α emissions to an isotope of lead, which by various β and α emissions yields another, stable, isotope of lead $^{206}_{82}\text{Pb}$. For any decay or radioactive transmutation the A and Z numbers of the daughter nuclei may be determined if the radiation emitted is known.

α emission: an α particle is two neutrons and two protons. For a parent nuclide $^{A}_{Z}\text{X}$ the daughter will be $^{A-4}_{Z-2}\text{X}'$.

β emission: a β particle is an electron emitted when a neutron changes into a proton. For a parent nuclide $^{A}_{Z}\text{X}$ the daughter will be $^{A}_{Z+1}\text{X}'$.

γ emission: the parent and daughter nuclides will be identical in symbol, $^{A}_{Z}\text{X}$, but the energy state of the daughter will be lower.

Measurement of half lives

The rate of disintegration of a radioactive specimen is experimentally determined at certain time intervals. The background count is also determined. A graph of activity due to the specimen against time is then plotted, which will have an exponential shape. From the graph the time taken for the activity to fall to half an initial value may be determined.

Example: From the graph in Figure 7.21 the counts per second will be the experimental readings minus the background count. The half life $T_{\frac{1}{2}}$ will be the time taken for the count

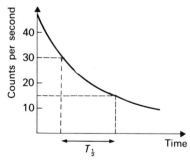

Figure 7.21 Determination of half life $T_{\frac{1}{2}}$

rate to fall from 30 to 15, as shown in the graph, or may equally be the time taken to fall from 40 to 20 counts per second, both giving the same result.

Note: The background count must always be subtracted from experimental readings before a graph is plotted.

Factors affecting the design of experiments to measure half lives $T_{\frac{1}{2}}$

If the daughter nuclides are radioactive, care must be taken to measure only the activity due to the parent nuclides. This may be achieved if:

(1) The daughter product is removed by chemical combination.
(2) The half life of the daughter product is much longer than that of the parent so that the activity due to the daughter may be neglected.
(3) The daughter emits a different type of radiation so that either an instrument sensitive to the parent's radiation may be chosen, or if the daughter's radiations are more easily absorbed they may be shielded from the detection apparatus.

Examples:

$$^{216}_{84}Po \rightarrow ^{212}_{82}Pb + ^{4}_{2}\alpha \qquad \text{half life} = 0.2\,s$$
$$^{212}_{82}Pb \rightarrow ^{212}_{83}Bi + ^{0}_{-1}\beta \qquad \text{half life} = 10\,\text{hours}$$

An ionization chamber more sensitive to α than β particles may be used to measure $T_{\frac{1}{2}}$ of $^{216}_{84}Po$. In addition, the half life of the daughter is much longer and therefore the number of α particle events is so much larger than any β particle events that the latter may be neglected.

To measure $T_{\frac{1}{2}}$ of the $^{212}_{82}Pb$ a paper shield may be used to absorb the α particles, or a scintillation counter with a β-sensitive phosphor may be chosen as a detector.

Note: When the half lives are measured the intensity or energy of the radiations emitted are not considered.

7:B.5   *Uses of Radioactivity*

(1) *Radioactive dating*

If the half life of the radioisotope is known, the ratio of parent to daughter nuclides in a rock may yield the time elapsed since the rock was laid down or the time passed since an animal or plant died. The isotope chosen must have a half life comparable to the age of the object being studied. For example, to determine the age of ancient rocks $^{238}_{92}U$, with a half life of 4500×10^6 years, is used, but for fossils and archaeological specimens $^{14}_{6}C$, with a half life of 5730 years is employed.

(2) *Tracer techniques*

A suitable radioactive isotope is introduced into the biological or chemical system. The subsequent motion of that nuclide may be followed by a detection instrument.

(3) *Radiotherapy*

$^{60}_{27}Co$ with its γ-rays is used to treat cancer patients. Radium needles are implanted in some cancers.

(4) *Industrial uses*

The ionization produced by α and β radioactive sources prevents the build-up of static charge hence the accumulation of dirt. Radiation may kill germs and may be used for sterilization. Reflection, refraction or scattering of radiation beams may detect flaws in materials. Thicknesses may be measured by noting the reduction in count rate from a β source.

(5) *Nuclear power*

This may be generated from suitable isotopes.

Problems

7.8 Why is Thomson's model of the atom inadequate to explain Rutherford's scattering experiment?

7.9 Which of the following nuclides are isotopes or isobars? $^{40}_{19}K, ^{40}_{18}Ar, ^{206}_{82}Pb, ^{207}_{82}Pb, ^{208}_{82}Pb$.

7.10 Which of the following may be deflected by an electric field: α particles, neutrons, γ-rays, β particles?

7.11 From a sample of uranium which is radioactive, 100 g of uranium chloride and 100 g of uranium nitrate were prepared. How will their half lives differ?

7.12 A radioactive nuclide has a half life of 2 hours. What percentage of its original activity remains after 8 hours?

7.13 The activity from a radioactive source was measured every day and the following results were obtained:

Time in days	0	1	2	3	4	5
Activity	137	92	63	44	31	22

The background count was found to be 7. Plot a graph and determine the half life of this source.

C: *Mass-energy Equivalence and Nuclear Reactions*

7:C.1 *Mass-energy Equivalence*

Mass can be considered as a form of energy
Einstein's equation relating mass m to energy E is

$$E = mc^2$$

where c is the velocity of light in a vacuum. Mass and energy are *jointly* conserved.

In some nuclear transformations it is observed that the total mass of the products does *not* equal the mass of the initial nuclides.

Example: Determine the mass difference and energy released in β decay. In β emission a neutron gives rise to a proton and a β particle

$$^1_0n \rightarrow \quad ^1_1p + ^{\ 0}_{-1}e$$

$$\text{Rest mass} \quad 16\,749 \quad 16\,726 \quad 9 \quad \times 10^{-31}\,kg$$

Total mass of left-hand side $= 16\,749 \times 10^{-31}\,kg$
Total mass of right-hand side $= 16\,735 \times 10^{-31}\,kg$
Mass difference $= 14 \times 10^{-31}\,kg$

Using Einstein's equation,

$$\text{Energy due to this mass difference} = 14 \times 10^{-31} \times 9 \times 10^{16}$$
$$= 1.26 \times 10^{-13} \, \text{J}$$

Most of this energy appears as the kinetic energy of the very fast moving β particles but a small amount is accounted for by nuclear recoil and heat.

Note: The law of conservation of mass does not apply but the law of conservation of energy still holds when the equivalence of mass and energy is taken into account.

Binding energy

It is observed that the mass of a nuclide is not equal to the total mass of its constituent neutrons and protons.

Example: Compare the rest mass of the deuterium nucleus with the mass of one neutron and one proton.

	^1_0n	^1_1p	^2_1H	
Rest mass	1.6749	1.6726	3.3436	$\times 10^{-27} \, \text{kg}$

Mass of $n + p = 3.3475 \times 10^{-27} \, \text{kg}$

Mass difference between $n + p$ and deuterium $= 0.0039 \times 10^{-27} \, \text{kg}$

The mass of the deuterium nucleus is *less* than the mass of one neutron and one proton.

Note: For any nuclide the mass of the nuclide is *less* than the total mass of the separate nucleons of which it is composed. This difference in mass is called the **mass defect**.

To take a nucleus completely apart, energy is required since there is a strong *nuclear force* of attraction between the nucleons.

The **binding energy** of a nucleus is the amount of energy required to remove all the nucleons to a large distance from each other. Thus the binding energy is the amount of energy equivalent to the mass defect.

$$\text{Binding energy} = (\text{total mass of nucleons} - \text{mass of nuclide}) \times c^2$$

From the above example the binding energy of deuterium is $3.5 \times 10^{-13} \, \text{J}$.

Example: Determine the binding energy of oxygen $^{16}_8O$ if the mass of one oxygen nucleus is $26.557 \times 10^{-27} \, \text{kg}$.

$$\text{Mass of } 8n + 8p = 26.780 \times 10^{-27} \, \text{kg}$$
$$\text{Mass defect} = 0.223 \times 10^{-27} \, \text{kg}$$
$$\text{Binding energy} = 2.007 \times 10^{-11} \, \text{J}$$

An interesting quantity is the *mean binding energy per nucleon*. For oxygen this is one sixteenth of the above energy, that is $1.25 \times 10^{-12} \, \text{J}$.

From the previous example the binding energy per nucleon of deuterium is only $1.75 \times 10^{-13} \, \text{J}$.

The *mean binding energy per nucleon* of a nuclide gives an indication of the stability of the nuclide. The larger its value the more stable the nuclide. If the mean binding energy *per nucleon* is large, this means that a large amount of energy is required *per nucleon* to pull the nuclide into its constituent nucleons.

Hence any change which increases the binding energy per nucleon will be accompanied by a *release* in energy.

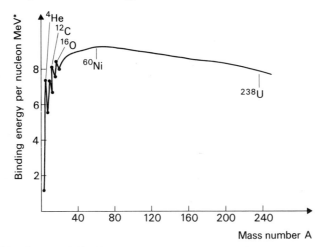

Figure 7.22 Graph of binding energy per nucleon against mass number A. The binding energy per nucleon reaches a maximum around A = 65 then declines towards the larger nuclei. Notice the position of the nuclei $^4_2$He, $^{16}_8$O, and $^{12}_6$C. These nuclei are particularly stable.

7:C.2 Nuclear Reactions

Nuclear fission

Nuclear fission is the splitting of a nuclide into smaller nuclides or particles. When some large nuclides are split into fragments it is observed that the mass of the products is *less* than the mass of the starting nuclides, thus again energy is emitted.

Example: Uranium-238 decays into thorium emitting α particles

$$^{238}_{92}U \rightarrow ^{234}_{90}Th + ^4_2He + 6.7 \times 10^{-13} J$$

The mass of the thorium plus the α particle is *less* than the mass of the uranium. The difference in mass appears as energy. Also the binding energy *per nucleon* of the thorium and helium is greater than that of the uranium. Hence the neutrons and protons are in a more stable state as thorium and helium than as uranium. Thus the larger nucleus will break up into the smaller ones with release of energy.

Example: If uranium-235 captures a slow neutron it may split into two smaller nuclides with a net gain in energy.

$$^1_0n + ^{235}_{92}U \rightarrow ^{92}_{36}Kr + ^{141}_{56}Ba + 3\,^1_0n + energy$$
$$(mass\ ^{92}_{36}Kr + mass\ ^{141}_{56}Ba + mass\ 3\,^1_0n)\ is\ less\ than\ (mass\ ^{235}_{92}U + mass\ ^1_0n)$$

Spontaneous fission of this nuclide of uranium also occurs but the above artificial transformation yields higher energies. Observe from the graph the higher binding energy per nucleon of the krypton and barium compared to uranium.

---

* The unit MeV is a unit of energy, 1 MeV = 1.602×10^{-13} J. For details see page 247.

Note: In **fission**, energy is released when a large nuclide breaks into smaller nuclides. The total mass of the products is less than the mass of the initial nuclide. This mass difference is released as energy. Fission may be spontaneous or induced.

Nuclear fission occurring under *controlled* conditions provides **atomic power** which may be converted into electricity. It is important to realize the large amounts of energy involved. The examples give the energy released *per fission*, i.e. for one nuclide. If one mole of uranium-235 was used, that is 235 g of uranium and 6×10^{23} atoms, the total energy released would be $2 \times 10^{-11} \times 6 \times 10^{23} = 1.2 \times 10^{13}$ J which is equivalent to a power of about 4 MW used *continuously* day and night for a *month*.

A chain reaction

To obtain energy from nuclear fission a sustained reaction is required. With uranium-235 if one neutron is captured the nuclide splits and three neutrons are released (see the equation on page 227). If these three were captured by another three uranium nuclei, nine more neutrons would then be available and so on, providing the possibility of a **chain reaction**. However, the neutrons emitted during fission are *fast* neutrons, i.e. they have high speeds, and these are most unlikely to cause further fission reactions. When slow neutrons, termed thermal neutrons, are captured by $^{235}_{92}U$, fission is much more likely to occur.

In a nuclear reactor, using uranium-235, neutrons are slowed down by placing a *moderator*, such as graphite, in the radioactive core. To *control* the chain reaction, cadmium or boron rods are inserted to capture excess neutrons when necessary.

Naturally occurring uranium has about 90% uranium-238 and the rest mainly uranium-235. The uranium-238 does *not* sustain a chain reaction, but $^{238}_{92}U$ can capture fast neutrons and produce plutonium. This plutonium can sustain a chain reaction.

$$^{1}_{0}n + ^{238}_{92}U \rightarrow ^{239}_{92}U \rightarrow ^{239}_{93}Np + ^{0}_{-1}e \quad \text{and} \quad ^{239}_{93}Np \rightarrow ^{239}_{94}Pu + ^{0}_{-1}e$$

In a **fast breeder reactor** natural uranium is enriched with either uranium-235 or plutonium-239 and surrounded with a covering of uranium. Many of the fast neutrons escape and are captured by the uranium-238 to produce plutonium. Thus the reactor can 'breed' more plutonium than the chain reaction is consuming.

The principle of the atomic bomb is the *uncontrolled* nuclear fission reaction of a certain critical sized piece of uranium or plutonium with the subsequent emission of a large amount of energy.

Nuclear fusion

Nuclear fusion is the building up of elements from smaller nuclides or protons and neutrons. The mass of the products may be *less* than the mass of the starting elements giving a net gain in energy.

Examples:

$$^{1}_{0}n + ^{1}_{1}p \rightarrow ^{2}_{1}H + \text{energy} \ (3.5 \times 10^{-13} \text{ J})$$
$$2^{2}_{1}H \rightarrow ^{3}_{1}H + ^{1}_{1}p + \text{energy} \ (6.5 \times 10^{-13} \text{ J})$$
$$^{3}_{1}H + ^{2}_{1}H \rightarrow ^{4}_{2}He + ^{1}_{0}n + \text{energy} \ (2.8 \times 10^{-12} \text{ J})$$

Observe that the binding energy per nucleon graph shows helium to be more stable than either deuterium or tritium. Some very high energies are available from these reactions.

Although high energies may be obtained from these nuclear fusion reactions, to initiate the reaction the two nuclides must *collide*. In many cases the starting nuclides are both positive and repel each other. However, if the temperature is considerably increased their mean velocities will increase and more collisions will take place enabling the reaction to proceed. These **thermonuclear** reactions occur naturally in the stars where the temperatures are very high, the temperature on the Sun being around 6000°C. This principle is also used in the hydrogen bomb where the detonator must provide the high temperature.

Note: In **fusion**, energy is released when small nuclides are built up into larger nuclides. The total mass of the larger nuclide and fragments formed is *less* than the starting nuclides. This mass difference is released as energy.

Observe that for elements with atomic numbers around 30 to 80 there is usually little gain for either fission or fusion.

7:C.3 Nuclear Particles

Artificial transformations

Certain radioactive nuclides occur naturally. Artificial radionuclides may be manufactured by bombarding stable nuclides with small nuclear particles such as α or β particles or neutrons.

Examples:

(1) $$^{7}_{3}\text{Li} + ^{1}_{1}\text{H} \rightarrow ^{4}_{2}\text{He} + ^{4}_{2}\text{He}$$

(2) $$^{27}_{13}\text{Al} + ^{1}_{0}\text{n} \rightarrow ^{24}_{11}\text{Na} + ^{4}_{2}\text{He}$$

the daughter $^{24}_{11}\text{Na}$ decays spontaneously (β particle)

$$^{24}_{11}\text{Na} \rightarrow ^{24}_{12}\text{Mg} + ^{0}_{-1}\text{e}$$

(3) $$^{27}_{13}\text{Al} + ^{4}_{2}\text{He} \rightarrow ^{30}_{15}\text{P} + ^{1}_{0}\text{n}$$

the daughter also decays $$^{30}_{15}\text{P} \rightarrow ^{30}_{14}\text{Si} + ^{0}_{+1}\text{e}$$ (positron)

The bombardment may cause a nuclide to be split, as in Example (1) or produce a nuclide of a higher atomic number as in Example (3). Also notice that the product depends on the bombarding particle; in (2) and (3) the same isotope of aluminium is used.

In these artificial transformations particles other than α and β may be emitted. In example (3) a neutron then a positive electron, the *positron*, are emitted which do not usually occur in natural radioactive decay. In nuclear fission and fusion artificial transformations are often involved as these may yield higher energies. In tracer techniques artificial isotopes are often required if the element has no naturally occurring radioisotope.

Nuclear particles emitted in nuclear reactions

α *and* β *particles* are emitted in both natural radioactive decay and artificial nuclear reactions.

The **neutron** was first discovered by Chadwick in 1932 when he bombarded beryllium with α particles.

$$^9_4\text{Be} + ^4_2\text{He} \rightarrow ^{12}_6\text{C} + ^1_0\text{n}$$

It has since been observed in many artificial nuclear reactions.

The **positron** is a positive electron with the same mass as the electron but a positive charge equal in magnitude to the electronic charge. The positron is the **antiparticle** to the electron and was first discovered in cosmic rays. Each particle has a corresponding antiparticle of equal mass but charge of opposite sign, e.g. the antiproton has a negative charge p^-.

Production of the positron

(1) In artificial nuclear reactions, when a proton is converted to a neutron.

$$^{30}_{15}\text{P} \rightarrow ^{30}_{14}\text{Si} + ^0_{+1}\text{e}$$

One proton of the phosphorus nucleus has been converted into a neutron giving a silicon nucleus.

(2) When a γ-ray is converted into an electron/positron pair.*

$$\gamma \rightarrow ^0_{-1}\text{e} + ^0_{+1}\text{e}$$

This is another example of mass–energy conversion.

$$\text{Mass of electron} = 9.1 \times 10^{-31}\,\text{kg} = \text{mass of positron}$$
$$\text{Mass of electron/positron pair} = 18.2 \times 10^{-31}\,\text{kg}$$
$$\text{Energy equivalent} = 18.2 \times 10^{-31} \times 9 \times 10^{16}\,\text{J}$$
$$= 1.64 \times 10^{-13}\,\text{J}$$

The energy of a photon of γ-rays is hf, where h is Planck's constant of 6.6×10^{-34} J s. Hence if the energy of the photon is converted into the rest mass of the electron/positron pair

$$1.64 \times 10^{-13} = 6.6 \times 10^{-34}\,f$$

giving the frequency f of approximately 2×10^{20} Hz, which is in the γ-ray range of the electromagnetic spectrum. This is the *minimum* frequency as the electron/positron pair are usually produced with high kinetic energies so that a higher frequency would be required in practice.

As momentum must be conserved the electron and positron will have equal velocities in opposite directions. They may be identified in a cloud chamber by applying a magnetic field perpendicular to the chamber and observing the curvature of their tracks, which will be equal but in opposite directions.

Conversely if an electron and positron combine the two γ-rays produced are termed the **annihilation radiation**.

$$\text{e}^+ + \text{e}^- \rightarrow 2\gamma$$

In a cloud chamber the electron and positron will leave tracks which will terminate where they meet and the γ-rays are produced.

---

*This example and the section on the neutrino are given for interest only. They are not required for examination purposes.

The neutrino

The **neutrino** and **anti-neutrino** are involved in positron emission and β decay to ensure there is conservation of **energy** and *angular momentum*. It is observed that the electrons emitted in β decay have a *range* of velocities, which implies that they are emitted with *different* kinetic energies. But, the example on page 225 indicates that the energy available is *constant* at 1.26×10^{-13} J. Hence another particle must be involved, which is also emitted with a variable amount of kinetic energy. The total energy of the *two* particles will be constant. Further, if the β particle has an angular 'spin' in one direction there must be another particle with an equal 'spin' in the complementary direction for the angular momentum to be conserved.

The complete reaction for β decay should be written:

$$_{1}^{0}n \longrightarrow {}_{1}^{+1}p + {}_{0}^{-1}e + \bar{v}$$

Neutrinos have zero mass (or a mass too small to detect). Notice that it is the anti-neutrino $\bar{v}$ which is involved in β decay, and the neutrino v in positron emission.

Fundamental particles*

A large number of other fundamental or elementary particles have been discovered in recent years.

There is a **strong** nuclear force, of very *short range*, between the proton and the neutron which binds these nucleons in the nucleus. In 1935 Yukawa postulated the existence of pions (π mesons) to account for this force, and these particles were later discovered in cosmic rays. The pions react strongly with the neutron and proton. There are three pions: π^-, π^0 and π^+.

$$\pi^- + p^+ \longrightarrow n + \pi^0 \quad \text{and} \quad \pi^+ + n \longrightarrow p^+ + \pi^0$$

The pions can also decay to produce muons (μ^-, μ^+) and mu neutrinos (v_μ).

$$\pi^+ \longrightarrow \mu^+ + v_\mu \quad \text{and} \quad \pi^- \longrightarrow \mu^- + \bar{v}_\mu$$

These decays are called **weak** interactions, (the strong nuclear force is not involved). Neutrinos, both v_e and v_μ, are uncharged particles which pass through matter, only very occasionally being involved in a reaction. Hence the term *weak* for interactions involving neutrinos, electrons, positrons, and muons. Also because the electron, positron, and muon are charged particles, they can be involved in **electromagnetic** interactions.

In high energy nuclear reactions and bombardments *many* more *particles* have been observed which involve strong interactions.

* This section is for interest only.

Nuclear particles can be classified into three groups:

(1) **Hadrons** (the *heavy* particles) which are involved in *strong* interactions.

Hadrons can be subdivided into two groups:

    (a) Baryons e.g. n, p^+, and many more heavier particles.

    (b) Mesons e.g. π^-, π^0, π^+ and many more of this type.

(2) **Leptons** (the *light* particles) which are involved in *weak* interactions; electron, positron, muons, and their neutrinos, plus a recently discovered tau lepton (τ).

(3) The **photon** γ, the particle of electromagnetic radiation, which is involved in all *electromagnetic* interactions.

Both the hadrons and leptons can have electromagnetic interactions.

So what is a fundamental particle? The atom is composed of a nucleus and electrons. The nucleus contains protons and neutrons.. The above large number of nuclear particles indicates that the neutron and proton are not 'fundamental' particles.

The present view is that the hadrons are composed of **quarks*** and antiquarks which have fractional charges of $\pm\frac{1}{3}$ and $\pm\frac{2}{3}$ of the electronic charge. A baryon is composed of 3 quarks but a meson is composed of a quark and an antiquark. However quarks *only* combine to form particles with an *integral charge*. The individual quarks have not been isolated. No doubt the future will hold more surprises on this subject.

Problems

7.14 Which of the following are conserved: momentum, energy, power, mass, charge? State clearly any conditions applicable.

7.15 Is the following radioactive transformation an example of nuclear fission, nuclear fusion or neither?

$$^{14}_{7}N + ^{4}_{2}He \rightarrow ^{17}_{8}O + ^{1}_{1}H$$

What is the name for the particle emitted?

7.16 The nucleus of lithium $^{6}_{3}Li$ has a mass of 9.9866×10^{-27} kg. Using the masses for the neutron and proton given on page 226, calculate the mass defect and binding energy of lithium.

7.17 Does nickel ^{60}Ni or lead ^{208}Pb have the larger (a) binding energy, (b) binding energy per nucleon?

7.18 State the main difficulty in producing nuclear fusion experimentally.

---

*The word quark appears as a nonsense word in *Finnegan's Wake* by James Joyce.

8 Interpretation of Experimental Data

A: Graphs

8:A.1 *Plotting a Graph*

In many experiments, the aim is to determine or verify a relationship between *two* physical quantities, when the remaining quantities are kept constant. A set of results is obtained, in mathematical language a set of ordered pairs. A graph can indicate *pictorially* how the two quantities are related.

The following list gives the points to remember when plotting a graph.
(1) A graph should have a **title**.
(2) Choose correctly the physical quantity to place on the *x* or *y* axis.
    (i) Plot the dependent variable on the *x*-axis, that is the quantity which is controlled during the experiment (the *cause*).
    (ii) Plot the independent quantity on the *y*-axis, that is the quantity which changes (the *effect*).
Remember if a question states: plot 'this' against 'that', e.g. current against frequency, the item mentioned first is plotted on the *y*-axis, i.e. current.
(3) **Label** the axes with the physical quantity and the unit.
(4) Choose a suitable **scale**. The graph should be large and fill the paper. Include the origin if necessary. Do not leave a large number of zeros with the numerical values, but place an appropriate power of ten with the unit. A graph relates *numbers*, thus if wavelengths ranging from 4×10^{-7} m to 7×10^{-7} m are used, the axes could be marked λ in 10^{-7} m, (or $\lambda/10^{-7}$ m).
(5) **Plot** points as $+$ or $\bigcirc$. Never use a dot alone which is easily lost.
(6) Draw a **smooth curve** or a **straight line** through the points, never join one point to the next. Never *assume* the graph passes through the origin. Non-coincidence with the origin may reveal an error.
(7) Plot the graph before clearing away the apparatus. Check any point which looks in error. Also, spread the readings over the whole available range.

Example: The current through a resistor is recorded for various applied potential differences V. Plot a graph of I against V.

Results:	p.d. *in volt*	1	2	3	4	5
	I *in ampere*	0.0019	0.0043	0.0063	0.0086	0.0111

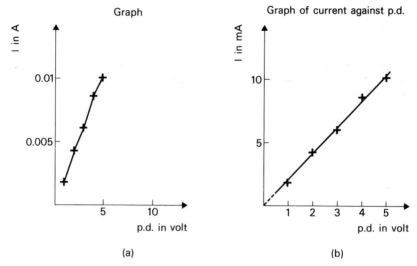

Figure 8.1 (a) Poor graph (b) correct graph

Graph (a) does not have a proper title, it has a poor scale on the *x*-axis, is badly labelled on the *y*-axis, and the points should not be joined.

8:A.2 *Graphical Verification of a Law or a Relationship*

In order to *verify* or determine the relationship between two physical quantities, a straight line graph must be obtained. For a *straight line* graph passing through the origin, (0,0) the quantity plotted on the *y*-axis varies directly as the quantity plotted on the *x*-axis.

$$y \propto x$$

Remember, a curve of any shape can only *suggest* a possible relationship. The shape of the curve can indicate the *function* of *x* which must be plotted in order to obtain the straight line, e.g. $1/x$, x^2, or $1/x^2$.

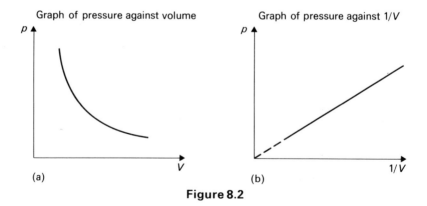

Figure 8.2

Example: The graph of the variation of the pressure of a gas p with the volume V, for a fixed mass of gas at a constant temperature, is given in Figure 8.2(a). What relationship does this graph suggest between p and V, and what further graph must be plotted in order to verify this relationship?

The graph of pressure against volume only *suggests* an inverse relationship. A further graph of p and $1/V$ should be plotted. This graph, Figure 8.2(b), *is* a straight line through the origin and this graph verifies that the relationship is $p \propto 1/V$.

Example: The distance s a trolley moves down a slope in a time t is recorded, and a graph of s against t is plotted, see Figure 8.3(a). What relationship does this graph suggest exists between s and t, and what further graph should be plotted?

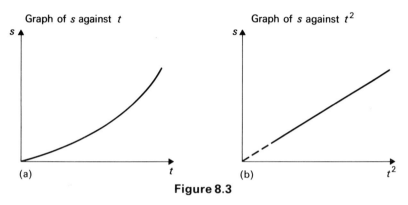

Graph of s against t Graph of s against t^2

(a) t (b) t^2

Figure 8.3

The graph in Figure 8.3(a) *suggests* a direct relationship of the form $y \propto x^2$ A graph of s and t^2 is then plotted, Figure 8.3(b). This graph *is* a straight line, which verifies that the relationship is $s \propto t^2$.

Example: The graph in Figure 8.4(a) was obtained from a set of results of the resistance R of 3 m lengths of nichrome wire of diameter d.

Results:

R (Ω)	44	29	7	5	3.6
d (mm)	0.28	0.35	0.71	0.82	1.00

What relationship does the graph of R against d suggest exists between R and d?

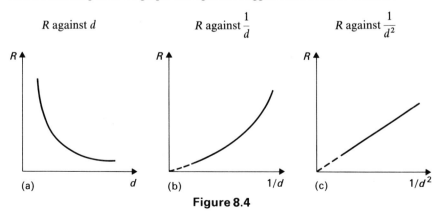

R against d R against $\dfrac{1}{d}$ R against $\dfrac{1}{d^2}$

(a) d (b) $1/d$ (c) $1/d^2$

Figure 8.4

The graph of Figure 8.4(a) *suggests* an inverse relationship. A graph of R and $1/d$, Figure 8.4(b) is then plotted, but this is *not* a straight line. The graph of R against $1/d^2$ *is a* straight line, hence the relationship is $R \propto 1/d^2$.

Note: In many cases an experiment is aimed at verifying a *known* law. Only the appropriate graph need be plotted and a straight line displayed. In the above example of resistance and diameter $R = \dfrac{\rho l}{A}$ giving $R \propto \dfrac{1}{A}$ or $R \propto \dfrac{1}{d^2}$. Only the last graph, Figure 8.4(c), is necessary to *verify* this relationship.

Preliminary inspection of a set of results

It would be very tedious to plot two or three graphs in order to verify a relationship. Simple inspection of the results can indicate if the relationship is 'direct' or 'inverse'. Further inspection can indicate if a direct relationship is of the form $y \propto x$ as opposed to $y \propto x^2$ or if an inverse relationship is of the form $y \propto 1/x$ or $y \propto 1/x^2$.

Note: It is important to remember that a physical relationship is required, not a mathematical discussion. For example, in a pendulum experiment, the period T^2 is plotted on the y-axis, see Figure 8.7, giving $T^2 \propto l$.

Determination of a relationship by the rate method *(calculation of the constant)*

After inspection of the results the constant of variation k should be calculated for the assumed relationship:

	(a)	(b)	(c)	(d)
Relationship	$y \propto x$	$y \propto x^2$	$y \propto \dfrac{1}{x}$	$y \propto \dfrac{1}{x^2}$
Equation	$y = kx$	$y = kx^2$	$y = \dfrac{k}{x}$	$y = \dfrac{k}{x^2}$
Constant k	$k = \dfrac{y}{x}$	$k = \dfrac{y}{x^2}$	$k = yx$	$k = yx^2$

For each pair of results, the constant k is calculated for an assumed relationship (a) or (b) or (c) or (d). If the values of the constant k are the same, within experimental error, it can be concluded that this is the required relationship.

Example: The heat H supplied in a given time interval to a given mass of water by an immersion heater, carrying a current I, is determined for different currents and the following results obtained:

$H (kJ)$	0.12	0.50	1.14	1.97	3.05
$I (A)$	0.5	1.0	1.5	2.0	2.5

From these results, obtain a relationship between H and I.

From inspection, the results suggest a direct relationship. First, consider is $y \propto x$? No, since the results $(0.50, 1.0)$ and $(1.97, 2.0)$ do not agree with this. Next consider whether $y \propto x^2$, i.e. is H/I^2 a constant? Calculate the values of k for $k = H/I^2$ for each pair of results.

$H\,(kJ)$	0.12	0.50	1.14	1.97	3.05
$k = \dfrac{H}{I^2}$	0.48	0.50	0.51	0.49	0.49

These values of k are almost the same, hence the relationship is $H \propto I^2$.

Note: In *all* problems, the constant of variation k should be calculated for at least *three* of the results and the correct relationship ascertained. Then, either the value of k is calculated and recorded for all the results, or if a graphical verification is required, *one* appropriate graph can be plotted and a straight line displayed.

(See Problems 4.2 page 113 and 5.11 page 157.)

8:A.3 *Information from the Gradient and Intercept of a Straight Line Graph*

Determination of the gradient and intercept of a graph can enable physical quantities to be evaluated.

The gradient and intercept of a straight line graph

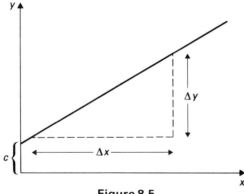

Figure 8.5

The equation of a straight line is

$$y = mx + c$$

The gradient m can be obtained from $\dfrac{\Delta y}{\Delta x}$.

The intercept c is the value on the y-axis when x is zero, that is the point where the graph cuts the y-axis and c can be positive or negative.

Note: To improve accuracy Δx and Δy should be made as large as possible. Their numerical values are obtained using the scales on the x and y axes.

Example: The resistance of a wire R is found to increase with the temperature T. The resistance is measured for a range of temperatures T and the following graph obtained, Figure 8.6. Determine from the graph:

(a) The resistance at 0°C.

(b) The rate of increase of resistance with temperature, that is the increase of resistance in ohms per kelvin.

(c) The temperature coefficient of resistance α, which is defined as the increase of resistance per kelvin divided by the resistance at 0°C.

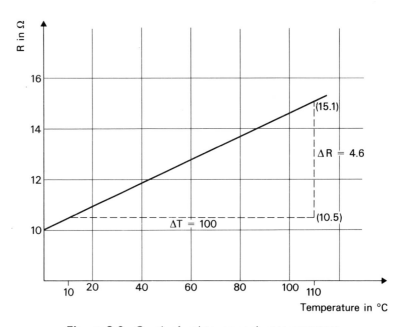

Figure 8.6 Graph of resistance against temperature

Note: The origin on the y-axis is not shown, the axes meet at the point (0,8).

(a) The resistance at 0°C, the intercept on the y-axis, is 10 Ω.

(b) The rate of increase of resistance with temperature, i.e. the gradient of the graph, is

$\dfrac{\Delta R}{\Delta T} = 0.046\,\Omega\,K^{-1}$. (Remember an interval of 1°C is the same as an interval of 1 K.)

(c) The temperature coefficient of resistance α is equal to $\dfrac{0.046}{10} = 4.6 \times 10^{-3}\,K^{-1}$.

(See also Problem 2.14 on page 68.)

Use of an intercept to reveal an error

A graph from an experiment may not pass through the origin due to experimental error. The reasons for the non-coincidence with the origin should then be considered.

Example: The period of a pendulum T was determined for different pendulum lengths l

and a graph of T^2 against l plotted. Theory gives the relationship $T^2 = \dfrac{4\pi^2 l}{g}$, hence a straight line through the origin is expected.

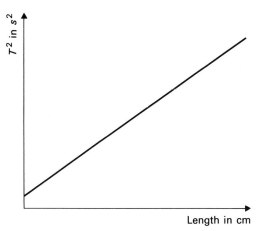

Figure 8.7 Graph of T^2 against l

The graph, which is obtained, does not pass through the origin, Figure 8.7. The measured lengths all appear to be too low. Perhaps an ordinary ruler was used which does not have the zero at its end? Perhaps all the measurements of length were made only to the top of the bob, not to the centre of the bob.

8:A.4 *Information from a Curve*

Radioactive decay curves

The half life can be determined from a graph of activity against time. These graphs are discussed in Chapter 7. No attempt is made at this level to replot a function and obtain a straight line.

The variation of activity from a β source with distance is also a curve of this type; both of these curves are exponential decay curves.

Other curves

An important use of a curve is to illustrate variations. For example, the variation in current when a capacitor is charged, or the build up of current in an inductive circuit, (see Chapter 4).

In some cases information is obtained from the maximum or minimum values. For example, the peak voltage can be obtained from a c.r.o. trace.

Example: An object is fired upwards at $54\,\mathrm{m\,s^{-1}}$ during a heavy hail storm. The following heights above the ground were determined. Plot a graph of height against time and determine the maximum height recorded and the time at which this maximum was reached.

height in m	48	85	124	118	100	72
time in s	1	2	4	6	7	8

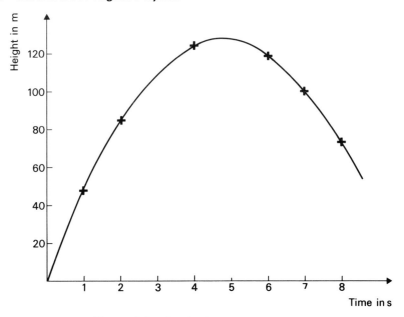

Figure 8.8   Graph of height against time

From the curve the maximum value of the height is 128 m at 4.7 s.

It is important to draw a *smooth* curve to obtain the value of the maximum. Never attempt to estimate a maximum from the readings. In this experiment, results at 3 s and 5 s would have been an advantage.

Problems

Note: The following problems from other Chapters are relevant: 2.14, 4.2 and 5.11.

8.1   The following results are obtained from a Boyle's law experiment.

Pressure in 10^4 Pa	10	11	13	16	18
Volume in cm	41	36	31	35	22

A student draws the following graph, Figure 8.9 and states that, 'this graph verifies Boyle's law since it is a straight line'.
(*a*) List all the reasons why this is a poor graph.
(*b*) Use the results and plot a correct graph to verify Boyle's law.

8.2   A pendulum is pulled to the side, a vertical height *h* above its rest position O. The pendulum is released and the speed *v* measured as it passes through this position O. The following results are obtained:

h *in cm*	1	3	5	7	9
v *in* ms^{-1}	0.44	0.77	1.00	1.17	1.33

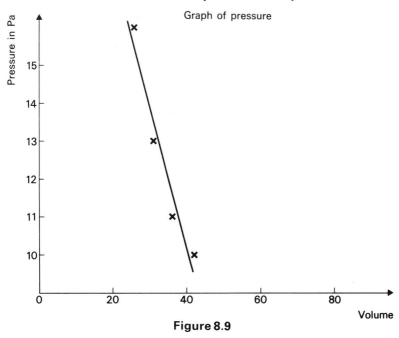

Figure 8.9

(*a*) Suggest a relationship between *h* and *v* using these results.

(*b*) Plot a suitable graph to verify this relationship. Remember to plot the independent variable on the *x*-axis.

(*c*) Use the principle of conservation of energy to derive a relationship between *h* and *v*, then use your graph to determine the acceleration due to gravity *g*.

8.3 The energy recorded on a joulemeter when a current is passed through a resistor *R* for a two minute period, is measured for different resistors. Using the same supply of e.m.f. *V*, and with negligible internal resistance, the following results were obtained.

R *in ohm*	10	20	30	40	50
Energy *in kJ*	1.73	0.86	0.58	0.43	0.34

(*a*) Use these results to illustrate graphically the relationship between energy and resistance.

(*b*) From your graph determine the e.m.f. of the supply.

8.4 A current *I* flows through a resistor *R* using a supply of e.m.f. ε and internal resistance *r*. The following results are obtained.

R *in ohm*	4	8	12	16	20
I *in A*	1.33	0.80	0.57	0.44	0.36

(*a*) Using the formula $\varepsilon = I(R+r)$ the relationship between *R* and *I* is $R = \dfrac{\varepsilon}{I} - r$.

Plot a graph of *R* against $\dfrac{1}{I}$ and obtain *E* and *r* from the graph.

(b) State the current which would flow if the external resistance R is equal to the internal resistance r.

8.5   In an experiment using a converging lens, a series of clear images were obtained on a screen for a range of object distances, and the following results obtained.

Object distance	u (cm)	20	25	30	40	50	60
Image distance	v (cm)	180	65	44	34	28	26

(a) Plot a graph of $\frac{1}{v}$ against $\frac{1}{u}$ and comment on its shape.

(b) The formula relating u, v, and the focal length f is $\frac{1}{f} = \frac{1}{u} + \frac{1}{v}$. Determine the focal length of the lens from your graph.

8.6   An inductor, capacitor and ammeter are connected in series with an alternating supply of variable frequency. The following results of the alternating current and frequency are obtained.

Current	in mA	10.2	17.2	28.9	27.7	18.6	13.4	10.5	8.7
Frequency	in Hz	90	110	130	150	170	190	210	230

(a) Plot a graph of current against frequency.
(b) State why the current varies with frequency.
(c) Obtain the maximum value of the current from your graph.
(d) At what frequency is the current a maximum?

B: Errors and Uncertainties

8:B.1   Sources of Experimental Error

The term error in this section does *not* include mistakes, such as setting up a circuit incorrectly or faulty arithmetic, due to individual incompetence.

Sources of experimental error can involve the design of the experiment.

Examples: If no attempt is made to keep the temperature constant, the measured value of a resistance will increase when a large current flows for a certain time.

In a specific heat capacity experiment heat exchange with the surroundings must be avoided. Therefore, either the experiment should be carried out in a container of poor thermal conductivity, or a control experiment set up to enable the heat exchanged with the surroundings to be determined.

The *effects* of sources of experimental error, on a measured value is important.

Example: A hot solid of known mass and temperature is added to a measured mass of liquid at room temperature, and the final temperature recorded. If no attempt is made to prevent any heat loss to the room, the final temperature measured will be too low. The specific heat capacity of the solid can be calculated using $m_s c_s (T_s - T) = m_d c_d (T - T_d)$, where the suffix s indicates the solid and d the liquid. Thus if T is too low the calculated value of c_s obtained will be too large.

Example:

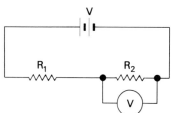

V is a d.c. supply of constant e.m.f.

Figure 8.10

The resistor R_1 is maintained in melting ice. What happens to the reading on the voltmeter if the supply is left on for some minutes? The resistance of R_1 will remain constant, but the resistance of R_2 will increase as its temperature rises. As the resistance of R_2 increases, the p.d. across R_2 will also rise. The voltmeter reading will *increase* with time.

Selection of the measuring instrument

A measuring instrument must be selected, which has the appropriate range and the required sensitivity. It is no use if the pointer is off the scale, or if all the readings are near to the zero of the scale.

Example: A metre rule is not suitable for the measurement of the diameter of a wire. A 0–1 A range is not a suitable ammeter for measuring currents between 2 mA and 4 mA.

8:B.2   *Uncertainties*

Even with sensitive equipment, a measured value will have some degree of inexactitude. The inexactitude may be termed either the *error* or the **uncertainty** in the value. (The term uncertainty will be used in this book.)

Random and systematic uncertainties

There are two kinds of uncertainties: random and systematic.
(1) **Random** uncertainties are the unpredictable variations in the result of identical measurements, caused by a lack of perfection in the observer, or the measuring instrument, or the act of measurement.
(2) **Systematic** uncertainties cause the results to be, either *all* too high, or *all* too low. These can be due to an instrument having a *zero error*, an instrument being incorrectly calibrated, or a *persistent* error on the part of the observer, e.g. starting a stop watch too early. In the pendulum graph example, Figure 8.7, there was a systematic uncertainty in the measurement of the length.

When a particular measurement is repeated a number of times, the random uncertainty is noticeable, but a systematic uncertainty is not revealed. The experimental conditions, instruments and observer need to be varied to attempt to correct for these systematic uncertainties. However, a graph can sometimes indicate the presence of a systematic uncertainty.

The uncertainty in a single value

Any particular value will have a random uncertainty, which is obtained using common sense.

Example: The uncertainty when *reading a scale:*
A metre rule can be read to 0.5 mm, hence a length could be (49.30 ± 0.05) cm.

Example: The uncertainty from a *setting*:
In a Wheatstone bridge circuit, there is no detectable movement of the galvanometer pointer when a variable resistor R is altered from $R = 31\,\Omega$ up to $R = 35\,\Omega$ giving $R = (33 \pm 2)\,\Omega$.

The **absolute uncertainty** is the actual amount of inexactness, i.e. 0.05 and 2 respectively, in the above examples.

The **fractional uncertainty** is the absolute uncertainty divided by the value, i.e. $\dfrac{0.05}{49.30}$ and $\dfrac{2}{33}$ respectively.

The **percentage uncertainty** is given by $\dfrac{\text{actual uncertainty}}{\text{value}} \times 100\,\%$, i.e. 0.1 % for the length measurement and 6.1 % for the resistance.

Note: The percentage (or fractional) uncertainty is important in ascertaining the accuracy of a measurement. The above example indicates that a more sensitive galvanometer is needed in the Wheatstone bridge circuit.

Treatment of uncertainties

An experiment can be designed in which a number of measurements are made and from these measurements, a final quantity calculated.

Examples: (*a*) The resistance at 0°C and the resistance at 100°C are measured and the change of resistance calculated. (*b*) The height and diameter of the base of a cylinder are measured in order to determine the volume of the cylinder.

The uncertainty in the *final* quantity or result must be determined from the uncertainties in the *individual* measurements using the following rules:
(1) For a **sum** or **difference**, add the absolute uncertainties to determine the absolute uncertainty of the result.
(2) For a **product** or a **quotient**, add the percentage (or fractional) uncertainties to obtain the percentage (or fractional) uncertainty of the result.

Examples: The resistance at 0°C is $(57 \pm 1)\,\Omega$ and at 100°C is $(71 \pm 1)\,\Omega$. The change of resistance is $(14 \pm 2)\,\Omega$. The absolute uncertainties are added.
The length of the sides of a triangle are (21.10 ± 0.05) cm, (14.15 ± 0.05) cm and (8.80 ± 0.05) cm. The perimeter is (44.05 ± 0.15) cm.

Example: The p.d. across a resistor R is (2.1 ± 0.1) V when a current of (82 ± 1) mA is flowing. The resistance is 25.6 Ω.
To determine the uncertainty, the *percentage* uncertainties must be *added*, i.e. 4.8 % + 1.2 % giving 6.0 %. The uncertainty in the result is 6.0 % of 25.6 which is 1.5 Ω. Thus the resistance is $(25.6 \pm 1.5)\,\Omega$.

Note: It is possible to use the limiting values to determine uncertainties, i.e. to evaluate the largest and smallest possible values, (see Answer to Problem 8.9 page 279).

Graphical representation of uncertainties

The uncertainties in *individual* experimental points can be shown on a graph by plotting the points as ⊢┼┤, or ⊡ or ⊢┼┤. The 'width' and 'height' of the points being determined by the uncertainties in the individual readings.

Example: The following results were obtained from an Ohm's law experiment. Plot a graph indicating the uncertainties in each value.

I *in mA*	1.2 ± 0.1	2.0 ± 0.1	3.1 ± 0.1
p.d. in V	0.9 ± 0.2	1.6 ± 0.2	2.6 ± 0.2

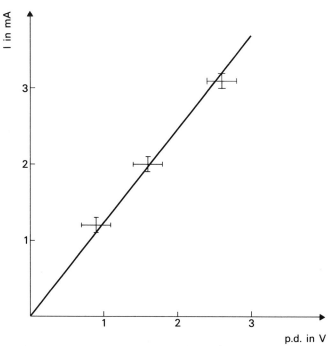

Figure 8.11 Graph of current against p.d.

The best straight line will now pass through all the points, because the uncertainty of each point is indicated on the graph.

8:B.3 *Heisenberg's Uncertainty Principle**

When measuring a physical quantity, the measuring instrument will disturb the system. Even in an ideal experiment, there will be some disturbance.

If a beam of γ-rays is directed at an electron to determine its position, Compton scattering could occur and the momentum of the electron would be altered. The use of electromagnetic radiation with a longer wavelength might

* Section 8:B.3 is not required for examination purposes.

cause less change in the momentum but, due to diffraction, the position would be less accurately measured, see Chapter 7. Quantum theory indicates that the disturbance made by an observer, when attempting to measure the position and momentum of an object are unpredictable, and *cannot* be accounted for *exactly*, however refined the apparatus.

The uncertainty principle states that the uncertainty in the position measurement Δx and the uncertainty in momentum measurement Δp_x in this x direction are such that their product is always larger than $\dfrac{h}{4\pi}$

$$\Delta x \, \Delta p_x \geqslant \frac{h}{4\pi} \qquad \text{where } h \text{ is Planck's constant.}$$

Notice that $h/4\pi$ is a *very small* value.

Thus even an idealized experiment performed perfectly will still give rise to some uncertainty, because of the quantum nature, or discontinuous nature, of matter.

Problems

8.7   A current measurement is (3.5 ± 0.1) mA. State the absolute and percentage uncertainties in this value.

8.8   The reactance of a capacitor is given by the formula $X_c = \dfrac{1}{2\pi f C}$. The capacitance of a capacitor is $(32\pm3)\,\mu$F and the frequency of a supply is (110 ± 2) Hz. Calculate the reactance at this frequency and state the uncertainty in the value.

8.9   In a metre bridge experiment to determine the resistance of a resistor X, the following results are obtained, (X is placed in the left arm of the bridge).
Length from left hand end to the balance point $b_1 = (45.0\pm1.0)$ cm.
Length from right hand end to the balance point $b_2 = (55.0\pm1.0)$ cm.
Standard resistor in the right arm, $S = (110\pm1)\,\Omega$.
Determine the resistance of X and the uncertainty in this value.

8.10   In an experiment to determine the power output of a heater, the temperature of (200 ± 1) g of water rose from 15°C to 26°C in 2 minutes. The thermometer could be read to ±0.5°C, and the stop clock to ±0.2 s.
(a) Determine the power output of the heater and the uncertainty in this value.
(b) Which measurement contributed most to the uncertainty?
(c) Comment on any points of experimental error in the design of this experiment, and state if these will tend to make the result too high or too low.

8.11

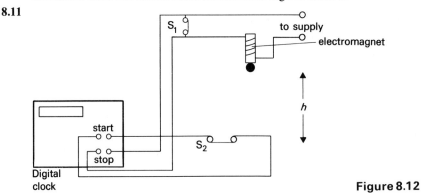

Figure 8.12

The steel ball is held by the electromagnet when S_1 is closed. When S_1 is opened the STOP circuit is broken and the clock commences to time until S_2 is opened. Thus the clock records the time t for a ball to fall freely from rest through a height h.

(a) Using $s = ut + \frac{1}{2}at^2$ with $u = 0$; $s = h$; $a = g$; then $h = \frac{1}{2}gt^2$ and $t^2 = \dfrac{2h}{g}$.

A set of results of the time t to fall different heights h is given below:

t in s	0.17	0.22	0.27	0.31	0.36	0.39
h in cm	8	16	25	36	50	60

Considering the above formula relating t and h, plot a graph with t on the y-axis and a suitable function of h on the x-axis to give a straight line.

(b) Explain why your graph suggests a systematic uncertainty in this experiment. Suggest a cause of this uncertainty.

(c) Determine g from the gradient of the graph.

(d) Does your value of g require any correction for the uncertainty discussed in (b)?

C: Units—Reference*

8:C.1 Non SI Units

The following units are not in the SI system of units, but may be encountered in certain texts.

The electronvolt (eV): *A unit of energy*
The electronvolt is the energy acquired by one electron accelerating through a potential difference of one volt.

$$1\,\text{eV} = 1.602 \times 10^{-19}\,\text{J} \quad \text{and} \quad 1\,\text{MeV} = 1.602 \times 10^{-13}\,\text{J}$$

This unit is widely used in nuclear physics, both as a unit of energy, and indirectly as a unit of mass. Since $E = mc^2$, a mass can be quoted in terms of energy.

Example: The mass of the electron is $9.1 \times 10^{-31}\,\text{kg}$ or $0.511\,\text{MeV}$.

$$0.511\,\text{MeV} = 0.511 \times 1.602 \times 10^{-13}\,\text{J} = 8.19 \times 10^{-14}\,\text{J}$$

Using $E = mc^2$, mass $= (8.19 \times 10^{-14})/(3 \times 10^8)^2 = 9.1 \times 10^{-31}\,\text{kg}$

The calorie: *A unit of energy*
The calorie is the heat required to raise the temperature of 1 g of water by 1°C.

$$1\,\text{calorie} = 4.18\,\text{J}.$$

The metre of mercury (m Hg): *A unit of pressure*
A metre of mercury is the pressure due to a one metre column of mercury, (see Section 6:B.2).

$$1\,\text{m Hg} = 133\,\text{kPa and } 1\,\text{mm Hg} = 133\,\text{Pa (Standard pressure} = 0.76\,\text{m Hg)}$$

* This section is for reference purposes only.

The atmosphere: *A unit of pressure*
The atmosphere is a useful unit for high pressures.

$$1 \text{ atmosphere} = \text{standard pressure} = 1.01325 \times 10^5 \text{ Pa.}$$

The bar: *A unit of pressure*
The bar, or millibar, is frequently used in weather forecasting.

$$1 \text{ bar} = 10^5 \text{ Pa and } 1 \text{ mbar} = 100 \text{ Pa (Standard pressure} = 1013.25 \text{ mbar)}$$

Notice that standard atmospheric pressure is a little larger than one bar.

8:C.2   *The SI System of Units*

The SI system of units is discussed in Section 1:A.1. The definitions of the **metre**, **second**, and **kilogram** are given on page 1. The definition of the **ampere** is on page 54, and that of the mole on page 185. For interest only, definitions of the **kelvin** and **candela** are given below.

One **kelvin** is the fraction, $\dfrac{1}{273.16}$, of the thermodynamic temperature of the triple point of water. (The triple point of water, 273.16 K, is the temperature at which steam, water, and ice may co-exist.)

The **candela** (cd) is defined as 1/600 000 of the luminous intensity per square metre of the surface of a black body, at the temperature of solidification of platinum, under 1.01325×10^5 Pa pressure.

Problem
(This problem is for interest only.)
8.12   The masses of the leptons are often quoted in MeV.

Particle	e^-	μ^-	τ^-
Mass (MeV)	0.511	105.7	1.9×10^3

Determine the mass of μ^- and τ^- in kg.

Answer to Problems

Chapter 1

1.1 Force: kg m s^{-2}, dimensions $[M][L][T]^{-2}$.

Power: $\dfrac{\text{energy}}{\text{time}}$ $\text{kg m}^2\,\text{s}^{-3}$, dimensions $[M][L]^2[T]^{-3}$.

Frequency: unit hertz, cycles per second, s^{-1}, dimensions $[T]^{-1}$.

Half-life: time taken for half the number of atoms in a sample to disintegrate, unit s, dimensions $[T]$.

Electric charge: unit coulomb, A s, dimensions $[I][T]$.

Pressure: $\dfrac{\text{force}}{\text{area}}$ $\text{kg m}^{-1}\,\text{s}^{-2}$, dimensions $[M][L]^{-1}[T]^{-2}$.

1.2 (a) The time taken for one slit to replace the previous one must be $\frac{1}{50}$ second. $\frac{1}{5}$ of a revolution takes $\frac{1}{50}$ second. Hence one revolution takes $\frac{1}{10}$ second. Number of revolutions per second $= 10$.

Or using $f = NR$ $50 = 5 \times R$, giving the rate of revolution $R = 10$ Hz.

(b) The highest single viewing frequency is 60 Hz. The strobe has 6 slits so that the motion is viewed six times every revolution of the disc. Frequency of the water waves $= 6 \times 60 = 360$ Hz.

Or using $f = NR$ for the highest single viewing frequency $f = 6 \times 60$ giving the frequency $= 360$ Hz.

1.3 (a) Using $nf = RN$ $3 \times 60 = 4 \times R$ $R = 45$ Hz.

(b) For the highest single viewing frequency $f = RN$ $60 = R \times 4$ giving $R = 15$ Hz. Single viewing is also observed when the disc is viewed every other cycle, i.e. $R = \frac{1}{2}15$, $R = 7.5$ Hz.

(c) At 16 Hz the white mark would not be 'stopped'. It would appear in a slightly different place each time.

first view

next view

clockwise rotation of disc

A single white line is observed *slowly moving* round in an anticlockwise direction.

1.4 $u = 0, v = 30\,\text{m s}^{-1}, t = 2\,\text{s}, a = ?, s = ?$.

Consider the acceleration first. Equation required is

$v = u + at$, $a = \frac{30}{2} = 15\,\text{m s}^{-2}$.

To determine s either of the other equations may be used.

Using $s = ut + \frac{1}{2}at^2$, $s = 0 + \frac{1}{2} \times 15 \times 4 = 30$ m.

1.5 From A to B the velocity is increasing therefore the object is accelerating. BC the velocity is constant. CD the velocity is decreasing, but CD is a curve hence the object has a non-uniform deceleration. (Notice that the object did not start from rest.)

1.6

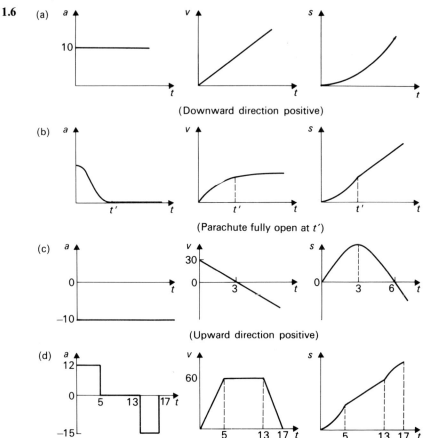

(a) (Downward direction positive)

(b) (Parachute fully open at *t'*)

(c) (Upward direction positive)

(d)

1.7 Remember that velocity is a vector quantity with direction, and speed is scalar. The initial *positive* direction will be called 'forwards'. The table below gives the answers to part (*a*) and (*b*) and the numerical values of *a* calculated from $a = \dfrac{v-u}{t}$ for part (*c*).

Section	OA	AB	BC	CD	DE
(*a*) speed	increasing	constant	decreasing	increasing	decreasing
(*b*) direction	forwards	forwards	forwards	reverse	reverse
(*c*) value of *a* in m s^{-2}	$\dfrac{12-0}{2}$		$\dfrac{0-12}{3}$	$\dfrac{-8-0}{2}$	$\dfrac{0-(-8)}{4}$
	$+6$	0	-4	-4	$+2$

Notice that for section CD, a is negative because the *direction* has reversed although the speed of the object increases. For section DE there are two negative factors giving a positive. The direction is reversed *and* the object is slowing down.

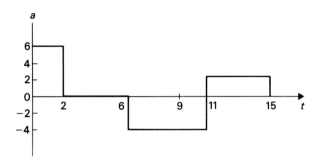

1.8 (a) Front of train (motion to signal box): $s = 150\,\mathrm{m}$, $a = 1\,\mathrm{m\,s^{-2}}$, $u = 0$, $v = $?.
Using $v^2 = u^2 + 2as$, $v = 10\sqrt{3} = 17.3\,\mathrm{m\,s^{-1}}$.
Back of train: $s = 200\,\mathrm{m}$, $a = 1\,\mathrm{m\,s^{-2}}$, $u = 0$, $v = $?.
Using $v^2 = u^2 + 2as$, $v = 20\,\mathrm{m\,s^{-1}}$.
(b) Time for the whole train to pass the signal: $a = 1\,\mathrm{m\,s^{-2}}$, $u = 17.3\,\mathrm{m\,s^{-1}}$, $v = 20\,\mathrm{m\,s^{-1}}$, $t = $?.
Using $v = u + at$, $t = 2.7\,\mathrm{s}$.

1.9 In the horizontal direction, velocity $= 40\cos 30$
$\qquad\qquad\qquad\qquad\qquad\quad = 34.6\,\mathrm{m\,s^{-1}}$
This horizontal velocity remains unchanged.
In the vertical direction the velocity $= 40\sin 30°$
$\qquad\qquad\qquad\qquad\qquad\qquad\quad = 20\,\mathrm{m\,s^{-1}}$
The vertical velocity, initially directed upwards, will decrease to zero, because of gravity, then increase in the downward direction. The initial vertical velocity is $20\,\mathrm{m\,s^{-1}}$ upwards, hence after $2\,\mathrm{s}$ the vertical velocity will be zero and the ball will start to fall. After the remaining $8\,\mathrm{s}$ the ball will acquire a downward velocity of $80\,\mathrm{m\,s^{-1}}$ just before hitting the rock.
(a)

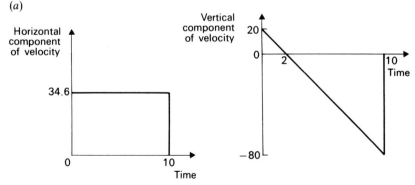

(b) The horizontal velocity remains constant.
Distance of the rock $= 34.6 \times 10 = 346\,\mathrm{m}$.
(c) Horizontal velocity just before landing $= 34.6\,\mathrm{m\,s^{-1}}$
Vertical velocity just before landing $= 80\,\mathrm{m\,s^{-1}}$

Resultant velocity   OR = 87.2

$$\tan \alpha = \frac{34.6}{80} \quad \alpha = 23.3°$$

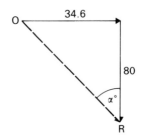

Velocity before landing = 87.2 m s^{-1} at an angle of 23.3° to the vertical (66.7° to the horizontal).

1.10   Horizontal distance travelled   = velocity × time
= 1000 × 40 = 4 × 10^4 m.
Vertical velocity when released   = $u = 0, a = +10$ m s$^{-2}, t = 40$ s, $s = $?.
Using $s = ut + \frac{1}{2}at^2$, $s = 0 + \frac{1}{2} \times 10 \times 16\,000$
Height of aircraft, $s = 8000$ m.
Vertical velocity before impact $v = $?.
$a = +10$ m s$^{-2}, t = 40$ s, $u = 0$.
Using $v = u + at$, $v = 10 \times 40 = 400$ m s^{-1}.
The resultant of the two velocities may be obtained by construction as shown in the Figure (b). (In this case it would be simple to calculate the final resultant velocity as a right-angled triangle is involved.)
Resultant velocity OR = 1080 m s^{-1}.

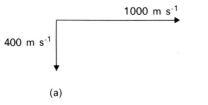

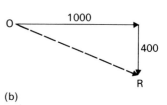

(a)   (b)

1.11   (a) The mean velocity between the 4th and 5th dot is $\dfrac{0.05}{1/50} = 2.5$ m s^{-1}.

The mean velocity between the 5th and 6th dot is $\dfrac{0.06}{1/50} = 3.0$ m s^{-1}.

Hence velocity at X = 2.75 m s^{-1}.

(b)   dots
velocities at the
midpoints between
successive dots
in m s^{-1} 1   1.5   2   2.5   3

The velocity increases by 0.5 m s^{-1} in every $\frac{1}{50}$ second. This is a uniform increase in velocity and so the acceleration is constant and $= \dfrac{0.5}{1/50} = 25$ m s^{-2}.

In practical experiments, once the acceleration has been checked to be constant, the velocity near the beginning and end are determined and the time interval between them calculated from the number of spaces. In the example above:

Initial velocity $= 1 \, \text{m s}^{-1}$ (midway between 1st and 2nd dot)
Final velocity $= 3 \, \text{m s}^{-1}$ (midway between 5th and 6th dot)
Time interval is the time taken for 4 spaces (NOT 5!)
$$= 4 \times \tfrac{1}{50} \, \text{s}$$

$$a = \frac{\text{change in velocity}}{\text{time taken}}$$

$$= \frac{3-1}{4/50}$$

$$= 25 \, \text{m s}^{-2}.$$

1.12

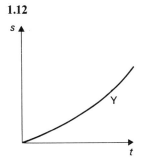 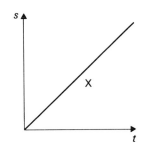 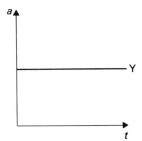

1.13 (*a*) At rest the balance will record the force on the mass due to the gravitational attraction of the Earth. Reading on balance $= 20 \, \text{N}$. ($g = 10 \, \text{m s}^{-2}$.)
(*b*) There are no extra *forces* on the mass when the lift moves with a constant velocity. Reading $= 20 \, \text{N}$.

 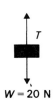

(*c*) The diagrams above are drawn and the forces are marked. *T* is the tension in the string holding the mass to the balance. By Newton's third law the reading on the balance equals *T*. From Newton's first and second laws the tension *T* will *not* equal the weight of the mass as the lift is accelerating upwards. **There must be an unbalanced force on the mass.**

Unbalanced upward force causing acceleration
$$= ma$$
$$T - 20 = 2 \times 2.5$$
$$T = 25$$
$$\text{Reading} = 25 \, \text{N}.$$

Note: If the lift were accelerating downwards the tension would be less than the weight and so the reading would be less than $20 \, \text{N}$.

1.14 Forces acting *on* the boy are shown below.

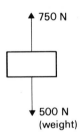

750 N

500 N
(weight)

Let upwards be positive
Using $F = ma$ $750 - 500 = 50a$
giving $a = 5\,\mathrm{m\,s^{-2}}$
The possible answers are: an upward acceleration of $5\,\mathrm{m\,s^{-2}}$
or a downward deceleration of $5\,\mathrm{m\,s^{-2}}$.
But if the lift starts *from rest* the only acceptable answer is an upward acceleration
of $5\,\mathrm{m\,s^{-2}}$.

1.15

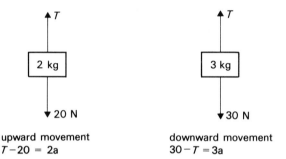

T

2 kg

20 N

upward movement
$T - 20 = 2a$

T

3 kg

30 N

downward movement
$30 - T = 3a$

(a) Add the equations $30 - 20 = 5a$ giving $a = 2\,\mathrm{m\,s^{-2}}$
(b) replacing this value of a in the first equation:
$$T - 20 = 2 \times 2 \text{ giving } T = 24\,\mathrm{N}$$

Notice that our common sense indicates the initial *directions* of movement when
the 2 kg mass is released.

1.16 (a)

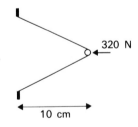

320 N

10 cm

$F = ma$ so $320 = 0.02 \times a$
$\therefore \quad a = 16\,000\,\mathrm{m\,s^{-2}}$.

(b) When the ball is about to leave the catapult the force on it is zero. When the
catapult is fully extended the force is 320 N.
Assuming the force is proportional to the extension.

$$\text{Average force} = \tfrac{1}{2}(320+0)$$
$$= 160\,\text{N}$$
$$\text{Average acceleration} = 8000\,\text{m s}^{-2}.$$

(c) Assuming the average acceleration acts over the entire 10 cm distance, $u = 0$, $s = 0.1\,\text{m}$, $a = 8000\,\text{m s}^{-2}$, $v = ?$.
Using $v^2 = u^2 + 2as$,

$$v^2 = 0 + 2 \times 8000 \times 0.1$$
$$= 1600$$
$$v = 40\,\text{m s}^{-1}.$$

(d) If the force is doubled the acceleration will double as $F = ma$, but it is the *square* of the velocity which doubles, $v^2 = 2as$ ($u = 0$).

$$\text{So } v^2 = 3200 \text{ and } v = 40\sqrt{2}\,\text{m s}^{-1}.$$

The velocity is increased by a factor of $\sqrt{2}$ when the force increases by a factor of 2.

1.17 (a) Initial momentum $= 2 \times 20 = 40\,\text{kg m s}^{-1}$.
(b) Impulse = change in momentum
$$= -8 - 40$$
$$= -48\,\text{kg m s}^{-1} \text{ or } -48\,\text{N s}.$$
(c) Impulse $= F\,\Delta t$. Force $F = \dfrac{-48}{3} = -16\,\text{N}$ (against the direction of motion of the ball).
(d) Using $F = ma$, $a = -8\,\text{m s}^{-2}$. $u = 20\,\text{m s}^{-1}$, $v = -4\,\text{m s}^{-1}$, $s = ?$.
Using $v^2 = u^2 + 2as$, $s = 24\,\text{m}$.
Observe that the final velocity is negative, $-4\,\text{m s}^{-1}$. Also notice that the ball travelled 25 m in the original direction and 1 m backwards giving a net displacement of 24 m.

1.18 (a) Component of the weight W down the plane
$$= W\cos 60° \text{ or } W\sin 30°$$
$$= 2 \times 10 \times \tfrac{1}{2} = 10\,\text{N}.$$
Resultant force on the trolley $= 10 - 4 = 6\,\text{N}$.

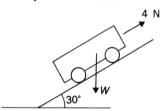

4 N

W

30°

(b) Using $F = ma$, $a = \tfrac{6}{2} = 3\,\text{m s}^{-2}$. $s = 6\,\text{m}$, $u = 0$, $v = ?$.
Using $v^2 = u^2 + 2as$, $v = 6\,\text{m s}^{-1}$.
(c) $F = 4\,\text{N}$ and, using $F = ma$, $a = 2\,\text{m s}^{-2}$. $u = 6\,\text{m s}^{-1}$, $v = 0$, $s = ?$.
Using $v^2 = u^2 + 2as$, $s = 9\,\text{m}$.

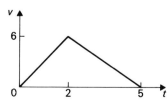

v

6

0   2   5 t

(*d*) Down the plane the time taken to acquire $6\,\mathrm{m\,s^{-1}}$ with an acceleration of $3\,\mathrm{m\,s^{-2}}$ is $2\,\mathrm{s}$. Along the plane the time taken to decelerate to rest is $3\,\mathrm{s}$.

1.19

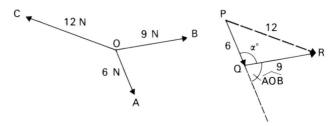

Because the forces are in equilibrium the vector addition of the 6 N and 9 N forces must equal 12 N.
The angle α may be determined either by construction using compasses or by using the cosine rule.
This gives $\alpha = 104.5°$ and angle AOB = 75.5°.

1.20 The acceleration due to gravity on this planet $= \frac{10}{4} = 2.5\,\mathrm{m\,s^{-2}}$.
Velocity after $6\,\mathrm{s} = 15\,\mathrm{m\,s^{-1}}$.
Increase in $E_k = \frac{1}{2} \times 3 \times 15^2 = 337.5\,\mathrm{J}$
Hence decrease in $E_p = 337.5\,\mathrm{J}$.

1.21 (*a*) Momentum is conserved. All the momentum of the bullet is transferred to the vehicle and the plasticine.

$$0.47 \times v = \frac{200 \times 0.12}{0.25} \qquad \text{leaving all the masses in g}$$

$$v = 204\,\mathrm{m\,s^{-1}}.$$

(*b*) Before being fired, the gun is at rest. Hence the total momentum after being fired is zero.

$$0 = (0.47 \times 10^{-3} \times 204) + 1.1 \times v_g$$
$$v_g = -0.087\,\mathrm{m\,s^{-1}}.$$

The negative sign indicates that the gun moves in the opposite direction to the bullet.

$$\text{Recoil velocity of the gun } v_g = 8.7\,\mathrm{cm\,s^{-1}}.$$

1.22 Vertical height $= 5 \times \sin 30° = 2.5\,\mathrm{m}$
Energy required to raise the block onto the platform (energy output)
    $= E_p$ to raise block vertically $2.5\,\mathrm{m} +$ work done against friction
    $= (60 \times 10 \times 2.5) + (11 \times 5)$
    $= 1555\,\mathrm{J}$

$$\text{efficiency} = \frac{80}{100} = \frac{\text{energy output}}{\text{energy input}} = \frac{1555}{P \times 3 \times 60}$$

Power rating $P = 10.8\,\mathrm{W}$.

1.23 (*a*)

distance h	0.5 m	1.5 m
$E_p = mgh$	10 J	30 J
$E_k = \frac{1}{2}mv^2$	4 J	24.01 J

(b) Energy lost due to friction is $E_p - E_k$ which equals 6.0 J, (two significant figures). This energy is converted into heat and sound.

(c) When $h = 1$, $E_p = 20$ J. Assuming the energy lost due to friction is 6 J the E_k available at the bottom is 14 J.

Using $E_k = \frac{1}{2}mv^2$, the velocity at the bottom $= 3.7 \, \text{m s}^{-1}$.

1.24 (a) Y is probably a softer ball since the time for which the bat is in contact with the ball is larger.

(b) The area under the graph represents the impulse on the ball or the change of momentum of the ball.

(c) Area under the graph for ball X $= \frac{1}{2} \times 4 \times 10^{-3} \times 2000 = 4$ N s.

$$\text{Change in momentum} = \text{mass} \times \text{change in velocity}.$$
$$4 = 0.1 \times (v - 0).$$

Giving the final speed $v = 40 \, \text{m s}^{-1}$.

(d) Since both the mass and the final speed of Y are the same as X, and their initial speeds are both zero, they both have the same change of momentum. Thus the area under both graphs must be equal to 4 N s. Q is less than 2000 and T is greater than 4 ms. Any pair of values, such as $Q = 1250$ N and $T = 6.4$ ms, can be chosen providing that $\frac{1}{2} \times Q \times T = 4$.

Note: The triangular shape of these graphs is only a simplified representation. In practice a *curve* would be more likely.

1.25 (a) Displacement (b) Change in velocity (c) Impulse or change of momentum (d) Energy or work done.

Chapter 2

2.1 Considering the energy changes:

(a) $eV = \frac{1}{2}mv^2$

$V = $ electric field $\times$ distance
$$= 3000 \times 0.05 = 150 \, \text{V}$$

$1.6 \times 10^{-19} \times 150 \qquad = \frac{1}{2} \times 9 \times 10^{-31} \times v^2$

$$v^2 = \frac{1.6 \times 150 \times 2 \times 10^{-19}}{9 \times 10^{-31}}$$

$$v = 7.3 \times 10^6 \, \text{m s}^{-1}$$

(b) If there is no electric field there will be no force on the electron hence its velocity will not change. Velocity at the screen $= 7.3 \times 10^6 \, \text{m s}^{-1}$.

(c) From cathode to anode the electron is accelerating.
$u = 0$, $v = 7.3 \times 10^6 \, \text{m s}^{-1}$, $s = 0.05$ m, $a = ?$, $t = ?$.

Notice that as $u = 0$, $v^2 = 2as$ and $v = at$, $v = \dfrac{2s}{t}$.

$$t = \frac{2 \times 0.05}{7.3 \times 10^6} = 1.36 \times 10^{-8} \, \text{s}.$$

From anode to screen the electron is travelling with a steady velocity so time

$$= \frac{\text{distance}}{\text{velocity}} = \frac{0.1}{7.3 \times 10^6} = 1.36 \times 10^{-8} \, \text{s}.$$

(d) The velocity given by $v^2 = \dfrac{2eV}{m}$ will be increased if the electric field is increased.

If the velocity increases the transit time will decrease when the distance between the plates remains constant.

2.2   One electron has a charge of 1.6×10^{-19} coulombs.

One coulomb contains $\dfrac{1}{1.6 \times 10^{-19}}$ electrons $= 6.25 \times 10^{18}$ electrons.

2.3   (a) The terminal velocity is measured to enable the *radius* of the drop to be calculated. From the radius the mass may subsequently be determined.
(b) A likely value is 1.6×10^{-19} coulombs.

result	3.18	12.80	4.79	6.41	8.01
no. of electronic charges	2	8	3	4	5

Assuming the results are accurate to one decimal place, a value of 0.8×10^{-19} is also consistent with this set of results but implies that the reading 12.80 has 16 electrons which is rather high.

2.4   (a)

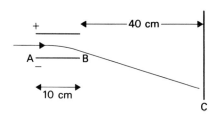

The proton accelerates inside the electric field AB, but the velocity is constant over the section BC.
(b) Horizontally: the time between the plates $= 10^{-6}$ s.
The horizontal component of the velocity is 10^5 m s^{-1} and will not change.

Vertically: Force $F = E \times Q = \dfrac{V}{d} + Q$

$$F = \dfrac{20}{0.04} \times 1.6 \times 10^{-19} = 8 \times 10^{-17}\,\text{N}.$$

Using $F = ma$, $a = 4.8 \times 10^{10}$ m s^{-2} (mass of proton $= 1.67 \times 10^{-27}$ kg).
Using $v = u + at$, $u = 0$, $a = 4.8 \times 10^{10}$ m s^{-2}, $t = 10^{-6}$ s
the vertical velocity on leaving the plates v is 4.8×10^4 m s^{-1}.
(c) After leaving the plates neither the horizontal or vertical components of the velocity will change, hence the velocity at the collector is the vector sum of these two velocities.

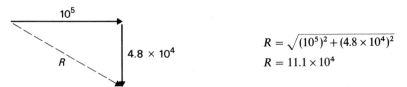

$R = \sqrt{(10^5)^2 + (4.8 \times 10^4)^2}$

$R = 11.1 \times 10^4$

Velocity at the collector $= 1.11 \times 10^5$ m s^{-1} at an angle of 25.6° to the horizontal.

2.5 (a) The forces on the drop are indicated below

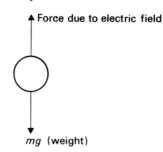

Force due to electric field

mg (weight)

(b) weight $= 1.44 \times 10^{-11} \times 10^{-3} \times 10 = 1.44 \times 10^{-13}$ N.
Since the drop is stationary the resultant force on the drop is zero.
Hence the force due to the electric field is 1.44×10^{-13} N, upwards.

(c) $\qquad F = \dfrac{V}{d} ne \quad$ since $\quad F = \mathbf{E} \times Q, \mathbf{E} = \dfrac{V}{d}$, and $Q = ne$

$$1.44 \times 10^{-13} = \left(\frac{100}{0.1 \times 10^{-2}} \right) \times n \times 1.6 \times 10^{-19}$$
$$n = 9.$$

There are nine electrons on the drop.

2.6 (a) p.d. across $1\frac{1}{2} \Omega$ resistor $=$ p.d. across 3Ω resistor
$\qquad\qquad\qquad\qquad\qquad = 3 \times 2 = 6$ V, using $V = IR$.
(b) current through $1\frac{1}{2} \Omega$ resistor $= 4$ A.
total current through 2Ω resistor $= 6$ A.
(c) 6 A is also the current through the source.
(d) lost volts $=$ current $\times$ internal resistance $= 6$ V.
e.m.f. of source, $\varepsilon = I(R + r)$
$\qquad\qquad$ Total $R = 2 + 1 + 1 = 4 \Omega$
$\qquad\qquad\qquad \varepsilon = 6(4 + 1) = 30$ V.

2.7 Electrical energy $= 2 \times 10^3 \times t$ J, where t is the time taken.
Heat required $= 0.5 \times 4.2 \times 10^3 \times 80$ J. 10% of the heat is lost, so

$\frac{90}{100} \times 2 \times 10^3 \times t = 40 \times 4.2 \times 10^3$

$$t = \frac{40 \times 4.2 \times 10^3 \times 100}{2 \times 10^3 \times 90}$$
$$= 93 \text{ s.}$$

2.8 Kilowatt hours used $= 60 \times 10^{-3} \times 2 = 120 \times 10^{-3}$
$\qquad\qquad\qquad$ Cost $= 0.12 \times 3 = 0.36$ p

Current $= \dfrac{\text{power}}{\text{p.d.}} = \dfrac{60}{240} = \frac{1}{4}$ A.

2.9 (a)
Total e.m.f. $= 3 \times 2.6 = 7.8$ V
Internal resistance $= 3 \Omega$
Using $\varepsilon = I(R + r)$,
$\qquad 7.8 = I(4 + 3)$
$\qquad\quad I = 1.1$ A

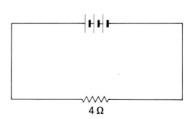

4Ω

(*b*)
Total e.m.f. = 2.6 V
Internal resistance is r.

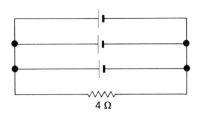

$$\frac{1}{r} = \tfrac{1}{1}+\tfrac{1}{1}+\tfrac{1}{1}; r = \tfrac{1}{3}\,\Omega$$

Using $\varepsilon = I(R+r)$,
$$2.6 = I(4+\tfrac{1}{3})$$
$$I = 0.6\,\text{A}$$

2.10   The p.d. across a particular component may not be calculated until the current has been determined.
Using $\varepsilon = I(R+r)$ for the whole circuit,
$$5 = I(8+2)$$
$$I = 0.5\,\text{A}$$
p.d. BC $= 4 \times 0.5 = 2\,\text{V}$ (using $V = IR$ between BC)
$$\begin{aligned}\text{t.p.d.} &= \text{e.m.f.} - \text{lost volts}\\ &= 5 - 2 \times 0.5\\ &= 4\,\text{V}\end{aligned}$$
Potential of C is zero; p.d. CD $= 3 \times 0.5 = 1\tfrac{1}{2}\,\text{V}$.
$\Rightarrow$ potential of D $= -1\tfrac{1}{2}\,\text{V}$.
Notice that the potential of A $= -1\tfrac{1}{2}+4 = +2\tfrac{1}{2}\,\text{V}$; potential of B $= +2\,\text{V}$.

2.11   The p.d. PQ is 8 V when XB is 20 cm.
(*a*) For PQ $= 4$ V then AX $= 10$ cm.
(*b*) For PQ $= 2$ V then XB $= 5$ cm and AX $= 15$ cm.
(*c*) For PQ $= 1$ V then XB $= 2.5$ cm and AX $= 17.5$ cm.
(*d*) There is a p.d. of 8 V for 20 cm, hence the potential gradient $= 0.4\,\text{V cm}^{-1}$.
(*e*) When a voltmeter is connected to PQ, the circuit has been altered. The total resistance between BX is *not* 50 Ω but the resistance of 50 Ω and 200 Ω in parallel.

Total resistance BX $= 40\,\Omega$ $\left(\text{Using } \dfrac{1}{R_{\text{T}}} = \dfrac{1}{R_1} + \dfrac{1}{R_2}\right)$.

Thus the 8 V is now across AX of 50 Ω and BX of 40 Ω.
Since AX has the larger resistance, the p.d. of AX will be more than 4 V

$$\text{The p.d. AX} = \tfrac{5}{9} \times 8 = 4.44\,\text{V}.$$

Note: When a circuit is altered all values of p.d. and current must be recalculated.

2.12

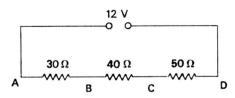

The three resistors are placed in series with the supply.
Possible values of the p.d. available are:

Between points	AB	BC	CD	AC	BD	AD
Value of p.d.	3 V	4 V	5 V	7 V	9 V	12 V

Observe that the total resistance is $120\,\Omega$ which gives a potential drop of 1 V per $10\,\Omega$ and a current of 0.1 A.

Note: This is one possible answer. If the resistors are connected in a different sequence, then other values of p.d. are obtained. For example, the $50\,\Omega$ resistor could be placed between the other two.

2.13 (a) Using $\varepsilon = I(R+r)$ then $12 = I(5+3+2)$ giving $I = 1.2\,A$
p.d. across the $3\,\Omega$ resistor $= 3 \times 1.2 = 3.6$ V
p.d. across the $5\,\Omega$ resistor $= 5 \times 1.2 = 6.0$ V.
(b) The rate of energy transfer, the power, is given by $P = IV$.
Rate of energy transfer in the $3\,\Omega$ resistor $= 3.6 \times 1.2 - 4.32$ W.
Rate of energy transfer in the $5\,\Omega$ resistor $= 6.0 \times 1.2 = 7.2$ W.
(c) The total power $= 12 \times 1.2 = 14.4$ W

$$\% \text{ transferred to the } 5\,\Omega \text{ resistor} = \frac{7.2}{14.4} \times 100 = 50\%.$$

2.14 (a) Using $\varepsilon = I(R+r)$ and $V = IR$ where V is the p.d. across R
$\varepsilon = V + Ir$
giving $\varepsilon - V = Ir$.
(b)

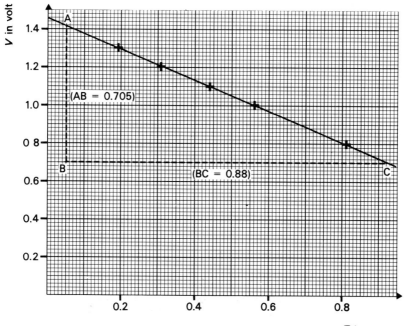

(c) The intercept on the y-axis is equal to ε, i.e. when $I = 0$ then $V = \varepsilon$.
From the graph $\varepsilon = 1.45$ V.
The gradient of the graph gives the value of r.
From the graph $r = \dfrac{AB}{BC} = 0.8\,\Omega.$

Observe the negative gradient of the graph and the negative sign with V in the equation.

(d) The resistor and ammeter are removed. The reading on the voltmeter then gives the e.m.f. The result will be slightly too low, since there would be some lost volts due to the small current flowing.

2.15 (a) With the ammeter recording the total current through R_x and the voltmeter, the current I is too high, using $R_x = \dfrac{V}{I}$ a value of R_x less than the correct value will be obtained.

(b) The p.d. reading is too high so the value calculated for R_x will be greater than its true value.

2.16 Yes, this method is correct. The second ammeter reading implies an equal current through R_x and the resistance box. Hence the two resistances are equal. As the second current is twice the first current the lost volts will be higher for the latter, so the p.d. across both R_x and the resistance box will decrease as the t.p.d. is less. This will introduce errors when the current is large, that is when R_x is small. In method (3) this error is eliminated.

Note: It is often useful to consider possible effects of the internal resistance of a cell or source, in questions which ask for sources of errors.

2.17 Using $\varepsilon = I(R+r)$, $1.5 = I(500+1)\,I = 0.003$ A.

$$\text{Lost volts} = 0.003 \text{ V.}$$
$$\text{True p.d. across voltmeter} = 1.5 - 0.003$$
$$= 1.497 \text{ V.}$$

This is the expected reading on the voltmeter, but the graduations are only every 0.1 V, therefore assuming a reading to half a graduation (that is to 0.05 V) may be estimated, the reading observed is 1.5 V. The conclusion drawn from this experiment is

$$\text{e.m.f. of cell} = 1.5 \pm 0.05 \text{ V.}$$

2.18 (a) $\varepsilon = I(R+r)$
  $20 = I(248+2)$
  $I = 0.08$ A. This is the current through the $200\,\Omega$ resistor.
p.d. across the $200\,\Omega$ resistor $= 0.08 \times 200 = 16$ V.
(b) With the ammeter in the circuit, the external resistance R has increased to $268\,\Omega$.
Using $\varepsilon = I(R+r)$ then $20 = I(268+2)$.
  $I = 0.074$ A. This is the current recorded by the ammeter.
(c) With the voltmeter in the circuit the external resistance will decrease.
  Total resistance of voltmeter and $200\,\Omega$ in parallel $= 150\,\Omega$.
Using $\varepsilon = I(R+r)$ then $20 = I(198+2)$ and $I = 0.1$ A.
Using $V = IR$ across the $200\,\Omega$ and voltmeter combination.
  $V = 0.1 \times 150$ for the total current and total resistance.
  Thus the p.d. recorded on the voltmeter $= 15$ V.
(d) The current recorded on the ammeter is less than the calculated current because the inclusion of the ammeter increases the overall resistance.

When the voltmeter is included the effective resistance of the $200\,\Omega$ resistor drops to $150\,\Omega$ and hence the p.d. decreases.

Note: The inclusion of an ammeter or a voltmeter alters the current in a circuit.

2.19 (*a*)

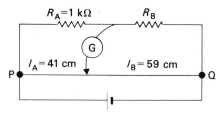

$R_A = 1\,\text{k}\Omega$ R_B

$l_A = 41$ cm $l_B = 59$ cm

P Q

Using $\dfrac{R_A}{R_B} = \dfrac{l_A}{l_B}$ gives $\dfrac{10^3}{R_B} = \dfrac{41}{59}$ and $R_B = 1.44\,\text{k}\Omega$.

With a 10% tolerance, the resistor R should have a value between $1.35\,\text{k}\Omega$ and $1.65\,\text{k}\Omega$.

Yes, the nominal $1.5\,\text{k}\Omega$ resistor is within its 10% tolerance.

(*b*) Resistance of $1.52\,\text{k}\Omega$ and $1.44\,\text{k}\Omega$ in parallel is $0.739\,\text{k}\Omega$.

Let the length to the balance point be x from P, and $(100-x)$ from Q.

$$\frac{10^3}{0.739 \times 10^3} = \frac{x}{(100-x)}$$

$$(100-x) = 0.739\,x$$

$$x = 57.5\,\text{cm}.$$

The balance point moves $(57.5-41) = 16.5\,\text{cm}$ to the right.

2.20 (*a*) $\dfrac{20}{80} = \dfrac{R}{120}$ giving $R = 30\,\Omega$.

(*b*)

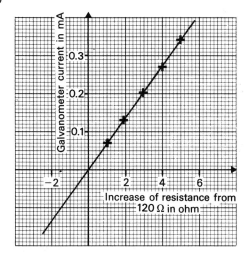

Galvanometer current in mA

0.3

0.2

0.1

−2 2 4 6

Increase of resistance from $120\,\Omega$ in ohm

The graph is a straight line through the origin, showing that:

(*change* of resistance from balance point) $\propto$ galvanometer current.

Chapter 3

3.1 No. An n-type semiconductor crystal contains small amounts of a *donor* not an acceptor. The crystal is *neutral* as the charges on the electrons are balanced by the positive charges on the nuclei.

3.2 (a) The transfer characteristic of the transistor is the graph of collector current obtained against base current, as the latter is altered.
(b) The current amplification factor is equal to

$$\frac{\text{change in collector current}}{\text{change in base current}}.$$

3.3 (a) When S is open, the transistor is off and the emitter/collector resistance is very large. There is no current flowing in the base circuit hence A_1 and V_1 will both be zero. There is no current, (or only a very small current), flowing in the collector circuit hence A_2 and V_2 are both zero. V_3 will display the value of V (supply).
(b) When S is closed the transistor conducts, and current will flow in both the input and output circuits. Since a transistor acts as an amplifier, the reading on A_2 will be larger than that on A_1. V_1 records the p.d. across R_1, and V_2 records the p.d. across R_2, Since I_2 is probably more than ten times I_1, the voltmeter V_2 will have the larger reading. The reading on V_3 will be almost zero, since the emitter/collector resistance is quite small when the transistor is conducting.

3.4 Peak current $= 3.1 \times \sqrt{2} = 4.38$ A.
The current drops to zero twice in each cycle, hence for a frequency of 50 Hz the current falls to zero 100 times in one second.

3.5 $f = \dfrac{1}{25 \times 10^{-3}} = 40$ Hz.

Heat produced per second $= \dfrac{I_m V_m}{2}$ 　　　　　　　　　　　　　$V_m = I_m R$

$$= \frac{30}{6} \times \frac{30}{2}$$ 　　　　　　　　　　　　　$I_m = \dfrac{30}{6}$

$$= 75 \text{ W}.$$

3.6 (a) Number of 1 cm squares for one cycle $= 8$.
Time for one cycle $= 8 \times 0.5 \quad = 4$ ms

$$\text{frequency} = \frac{1}{4 \times 10^{-3}} = 250 \text{ Hz}.$$

(b) Peak p.d. $= 3 \times 50 \times 10^{-3} = 0.15$ V.
$$V_{\text{r.m.s.}} = \frac{0.15}{\sqrt{2}} = 0.106 \text{ V}.$$

3.7 The frequency f is given by $f_X : f_Y = 1 : 2$.

So $\dfrac{f_X}{f_Y} = \frac{1}{2}$ and $\dfrac{80}{f} = \frac{1}{2}$, giving $f = 160$ Hz.

3.8 (a) Time of contact $= 0.2$ ms.
(b) Time for one cycle $= 6 \times 0.1$ ms

$$\text{frequency} = \frac{1}{6 \times 0.1 \times 10^{-3}} = 1.67 \text{ kHz}.$$

(c) From the heights of the trace, the p.d. across R_1 is three times that across R_2.
Hence R_1 is three times R_2
giving $R_2 = \frac{1}{3} \times 3.3 = 1.1\,k\Omega$.
(d) The p.d. across $R_1 = 6\,V$ (lower three squares).

Using $V = IR$ across R_1, the current $I = \dfrac{6}{3.3 \times 10^3} = 1.8\,mA$.

The current through $2.2\,k\Omega$ resistor is $1.8\,mA$.
The t.p.d. of the source = p.d. across R_2 + p.d. across R_1 + p.d. across $2.2\,k\Omega$

$$\text{t.p.d.} = 2 + 6 + 4 = 12\,V.$$

3.9 So that the electrons may travel to the anode without collisions on the way. If the envelope were not evacuated a considerably larger p.d. would be required to obtain any current flow.

3.10 No heat circuits are required; no warm up time; small size; cheaper, low operating voltage, low supply currents; less easily damaged.

Chapter 4

4.1 (a) Initially the capacitor has no charge and the rate of flow of charge is determined by the resistor.
Using $V = IR$ gives the current $I = 4\,mA$.
(b) When S is closed, the current through the resistor is $4\,mA$.
Hence the p.d. across R is $4\,V$.
When the capacitor is fully charged, no current will flow and the p.d. across R will be zero.
(c) When S is first closed, the p.d. across the capacitor is zero. When fully charged the p.d. across the capacitor is $4\,V$.

(d) Using $C = \dfrac{Q}{V}$ the final charge stored, $Q = 4 \times 1000 \times 10^{-6} = 4\,mC$.

Using energy $= \frac{1}{2}CV^2$, the energy stored $= 8\,mJ$.

4.2 (a)

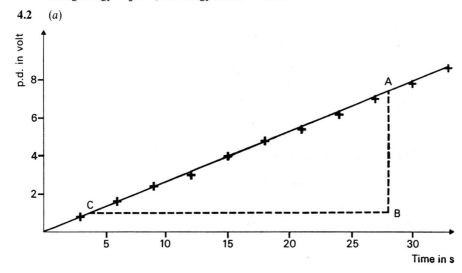

(b) The graph is a straight line through the origin hence $V \propto t$.

But $I = \dfrac{Q}{t}$ or $It = Q$.

Since the current I is a constant $t \propto Q$
giving $V \propto Q$ for constant current.

(c) The gradient of the graph $\dfrac{\Delta V}{\Delta t} = \dfrac{AB}{BC} = \dfrac{7.4 - 1.0}{28 - 4} = 0.267$.

But $C = \dfrac{Q}{V} = \dfrac{It}{V} = I \times \dfrac{1}{\text{gradient of graph}}$

$C = \dfrac{0.3 \times 10^{-3}}{0.267}$

$C = 1.12 \times 10^{-3}\,\text{F}$ (i.e. just over $1000\,\mu\text{F}$).

4.3 (a) Using $C = \dfrac{Q}{V}$,

$Q = 50 \times 10^{-6} \times 8$
$\quad = 400\,\mu\text{C}.$

(b) p.d. across C just before discharging $= 8\,\text{V}$ and $R = 32\,\Omega$.
Using $V = IR$, $I = 0.25\,\text{A}$.
(c) The current decreases from $0.25\,\text{A}$ to zero as the capacitor discharges.
(d) If the resistance increases the initial current would be less and the ammeter pointer would move more slowly to zero; the capacitor taking longer to discharge.

4.4 (a) Charge $=$ current $\times$ time
Q after $10\,\text{s} = 0.2 \times 10^{-3} \times 10$
$\qquad\qquad = 2 \times 10^{-3}\,\text{C}$
Q after $30\,\text{s} = 0.2 \times 10^{-3} \times 30$
$\qquad\qquad = 6 \times 10^{-3}\,\text{C}.$

Using $C = \dfrac{Q}{V}$ and the $10\,\text{s}$ result

$C = \dfrac{2 \times 10^{-3}}{1.6} = 1.2 \times 10^{-3}\,\text{F}$

with the $30\,\text{s}$ result

$C = \dfrac{6 \times 10^{-3}}{5} = 1.2 \times 10^{-3}\,\text{F}.$

Capacitance $= 1.2 \times 10^{-3}\,\text{F}$.
(b) Initially there is no charge on the capacitor and the maximum current will flow.
Using $V = IR$, $6 = 0.2 \times 10^{-3} \times R$ hence $R = 30\,\text{k}\Omega$.
As the capacitor charges up the value of R must be *decreased* in order to maintain a $0.2\,\text{mA}$ current.
Range of R must be 0–$30\,\text{k}\Omega$.

4.5 L_1 has the larger inductance value because B_2 took longer to light up.
L_1 has a core in position and has the same resistance as R because B_1 and B_2 are eventually of the same brightness.
L_2 has no core but probably has more turns than L_1 since its resistance is larger.
B_3 was dimmer.

4.6 (a) $\frac{1}{6} = \dfrac{x}{240}$, $x = 40$. Hence the secondary p.d. $= 40\,\text{V}$.

(b) If there is no power loss $240 \times 2 = 40 \times I$, $I = 12$ A. But there is power loss and therefore the secondary current will be *less* than 12 A.

(c) The core should be laminated with strips of non-conducting material. The answer to (a) will be the same, the turns ratio affecting the p.d. The answer to (b) will be different. The current will be nearer the 12 A but there may be other power losses and it is unlikely that the current will attain the full 12 A.

4.7

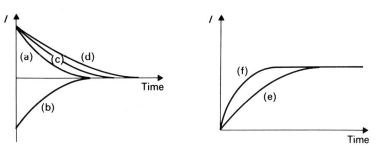

(a) The initial current dies away as the capacitor charges up.

(b) The current is in the opposite direction as the capacitor discharges.

(c) If the overlap of the plates is increased the capacitance increases and therefore the time for all the charge to flow off the plates increases.

(d) An increase in dielectric constant increases the capacitance hence the time for the current to decay is greater.

(e) With an inductor the current takes time to build up to a steady value due to the induced back e.m.f.

(f) The inductance decreases when the core is removed, hence the back e.m.f. decreases and the build-up time will be shorter.

4.8 As the frequency is increased, the opposition to the current X_C decreases, therefore the current will increase. If both switches are closed the two capacitors are in parallel. This results in a *larger* overall capacitance. A larger capacitance has a lower opposition X_C to the current and therefore higher current readings are obtained over the whole frequency range.

4.9 (a) If the current decreases the reactance must increase with frequency. Thus Z is an inductor.

(b) Z is a resistor. The resistance is independent of the frequency.

(c) Z is a capacitor. X_C decreases with frequency hence the current will increase with frequency.

(d) They refer to (c) since $X_C \propto \dfrac{1}{f}$ giving $I \propto f$ as shown by the graphs.

4.10 Because of the initial calibration the *height* of the trace on each c.r.o. indicates the *potential difference* across that component.

(a) At high frequencies X_L is much larger than X_C and therefore the p.d. across L will be larger, giving the trace shown below. Remember, $V = IX$ and the currents through L and C are the same. There will be no observable trace on c.r.o. 2, the p.d. across C being very small.

c.r.o. 1   c.r.o. 2

(b) At lower frequencies X_C is large and hence the p.d. across the capacitor will be much greater. Observe the trace on c.r.o. 2.

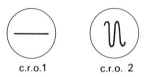

c.r.o.1   c.r.o. 2

4.11   (a) At low frequencies X_C is greater than X_L and more current will pass through L making B_1 brighter.
(b) At high frequencies X_L is larger and the easier current path is now through C. Bulb B_2 will be brighter.

4.12

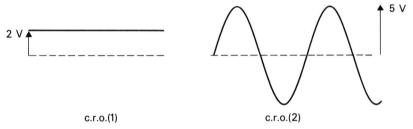

c.r.o.(1)   c.r.o.(2)

The 2 V direct supply will cause a current to flow until the capacitor is fully charged to a p.d. of 2 V. Then *no* more direct current will flow. The a.c. supply of 10 V will provide 5 V in the circuit. Both the resistor and the capacitor will allow an alternating current to flow. The reactance of the capacitor $\dfrac{1}{2\pi f C}$ is much smaller than the resistance of the resistor since C has a large value. Hence the alternating potential across $R \gg$ alternating p.d. across C. A c.r.o. displays the p.d. across that component, and c.r.o. (2) will show an alternating trace with almost 5 V peak value.

Chapter 5

5.1   The refractive index $= \dfrac{\text{velocity of light in vacuum}}{\text{velocity of light in the medium}}$

hence refractive indices will *always* be greater than unity.

5.2

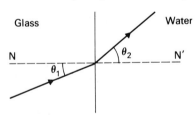

(a) The angle in glass will be *smaller* than the angle in water because glass has the *higher* refractive index.
(b) Notice that the angles of incidence are not given directly; θ_1 being the angle between the ray and the *normal*.

Using $n_1 \sin \theta_1 = n_2 \sin \theta_2$, and $n_1 = 1$ for air.
For perspex: $n = 1.5$, $\theta_1 = 90 - 30 = 60°$.
$1 \times \sin 60 = 1.5 \sin \theta_2$, hence $\theta_2 = 35.3°$.
Angle of refraction $= \theta_2 = 35.3°$.
For water: $1 \times \sin 50 = 1.33 \times \sin \theta_2$, hence $\theta_2 = 35.1°$.
For quartz crystal: $1 \times \sin 30 = 1.54 \sin \theta_2$, hence $\theta_2 = 18.9$.
For glycerol: $1 \times \sin 40 = 1.47 \times \sin \theta_2$, hence $\theta_2 = 25.9°$.

5.3 The frequency remains *unchanged* as the number of waves produced per second is determined by the *source* not the material.

The velocity in the glass v_g may be calculated from the relation $n = \dfrac{v_0}{v_g}$ where v_0 is the velocity of light in air.

$$1.5 = \frac{3 \times 10^8}{v_g},$$

hence
$$v_g = 2 \times 10^8 \text{ m s}^{-1}.$$

As would be expected, the velocity of the light in the glass is less than that in air.
The wavelength is given from $v = \lambda f$ where v is the velocity of light in glass and λ the wavelength in the glass.
Hence $2 \times 10^8 = \lambda \times 6 \times 10^{14}$, giving $\lambda = 3.3 = 10^{-7}$ m.

Note: Compare this with the wavelength of light of this frequency in air which is $\dfrac{3 \times 10^8}{6 \times 10^{14}}$ equal to 5×10^{-7} m.
This illustrates that when light enters a medium of higher refractive index the velocity *and* the wavelength decrease.

5.4 (a) Using $\dfrac{v_1}{v_2} = \dfrac{\sin \theta_1}{\sin \theta_2}$ then $\dfrac{2.26 \times 10^8}{1.7 \times 10^8} = \dfrac{\sin 50}{\sin \theta_2}$.

Giving θ_2, the angle in the glass $= 35.2°$.
(b) The wavelength is longer in water. Both the speed and wavelength decrease in the glass.
(c) The frequency, which depends on the source, is the same in both the water and the glass.
(d) Speed of light in air $= 3 \times 10^8$ m s^{-1}.
Using $\dfrac{v_1}{v_2} = \dfrac{\lambda_1}{\lambda_2}$ then $\dfrac{3 \times 10^8}{1.7 \times 10^8} = \dfrac{580 \times 10^{-9}}{\lambda_2}$.
Giving the wavelength in the glass $= 329$ nm.

5.5

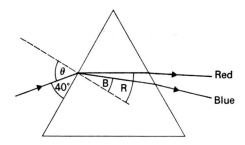

Angle of incidence θ is the angle between the ray and the normal. Hence $\theta = 90 - 40 = 50°$. Let the angles of refraction for blue and red light be B and R respectively.

For red light:
$$1.51 = \frac{\sin 50}{\sin R}$$

$$\sin R = \frac{\sin 50}{1.51} = 0.5072$$

$$R = 30.5°.$$

For blue light:
$$B = 30.5 - 0.7 = 29.8°$$

$$n_{blue} = \frac{\sin 50}{\sin 29.8} = 1.54.$$

Refractive index for blue light $= 1.54$.

5.6   A has the larger focal length. The refractive index of B is greater so light will be 'bent' more by this lens and be brought to a focus *nearer* the lens, hence its focal length is shorter.

5.7   Either. If the image was magnified a converging lens was used with the object nearer to the lens than its focal length. If the image was diminished the lens used was diverging for any object position.

5.8

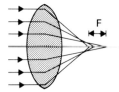

(*a*) A wide beam of light will give a blurred focus. This lens is thicker and so will have the shorter focal length.
(*b*) The light is directed nearer to the centre of this lens so a sharper focus is obtained.

5.9

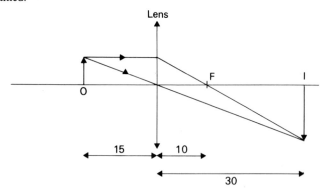

The image is magnified, real, and inverted. The projector must produce a real magnified image. Therefore the object distance must lie between f and $2f$, where f is the focal length. Thus the object distance is between 10 cm and 20 cm.

5.10

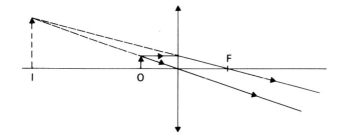

The image is magnified, virtual, and erect and 60 cm from the lens.

$\left(\text{Using } \dfrac{1}{f} = \dfrac{1}{u} + \dfrac{1}{v}, \dfrac{1}{20} = \dfrac{1}{15} + \dfrac{1}{v}, \text{ giving } v = -60 \text{ cm.} \right)$

The magnification is $\frac{60}{15} = 4$. Hence the width of the image = $3.5 \times 4 = 14$ cm. Only useful in a slide viewer. A slide projector throws a *real* magnified image on to a screen.

5.11

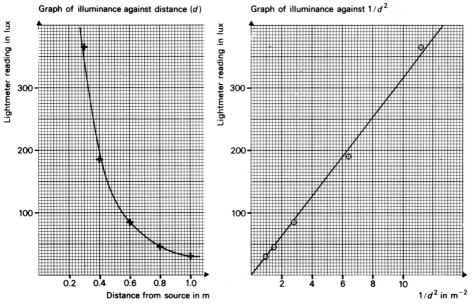

The second graph of illuminance and $\dfrac{1}{d^2}$ is a straight line through the origin showing $E \propto \dfrac{1}{d^2}$.

5.12 $\frac{1}{4}$ wavelength. The distance between consecutive maxima is $\frac{1}{2}\lambda$.

5.13 A change in the depth of the water. The 3 cm wavelength waves are in the shallow water. Relative change in velocity will be in the ratio 5:3 for A to B. There is no change in frequency.

5.14 Using $v = \lambda f$; for light $v = 3 \times 10^8 \, \text{m s}^{-1}$ hence $f = 6 \times 10^{14} \, \text{Hz}$
for sound $v = 340 \, \text{m s}^{-1}$ hence $f = 6.8 \times 10^8 \, \text{Hz}$.

5.15 (a) A standing wave is set up as the incident and reflected waves interfere. Hence the distance between consecutive minima will be $\frac{1}{2}$ the incident wavelength, that is 0.05 m.
(b) The two travelling waves interfere hence the distance between two minima is equal to the wavelength, that is 0.1 m.

5.16 (a)

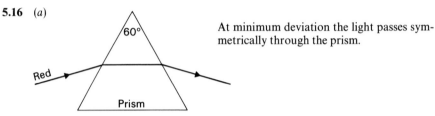

At minimum deviation the light passes symmetrically through the prism.

(b) For the prism, the blue will be deviated more than the red.

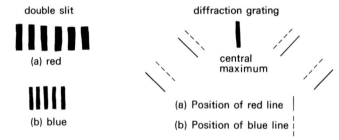

The patterns for the blue light are closer together than those for the red light. There is less deviation of the blue light. For the diffraction grating there are an *odd* number of coloured lines.
Compare the patterns of the single slit on page 161 with these of the double slit.

5.17 (a) The value of n must be an even integer. The path difference must be a whole number of wavelengths for a maximum.
(b) For a minimum n must be an odd integer.
(c) X must be equidistant from S_1 and S_2.
(d) There are five wavelengths in 22 cm hence $\lambda = 4.4 \, \text{cm}$.
The distance $S_1 P$ is 8 wavelengths, hence $S_2 P = 35.2 \, \text{cm}$.

5.18 (a) The distance across 8 minima will be eight wavelengths $= 16 \, \text{cm}$.
(b) Let this 8th minimum position be X.
The path difference $S_2 X - S_1 X = (n + \frac{1}{2})\lambda$ for minima only
$$S_2 X - 30 = (8 + \frac{1}{2})\lambda \quad n = 8$$
giving $S_2 X = 47 \, \text{cm}$.

(Using $S_2 X - S_1 X = \dfrac{n\lambda}{2}$ n will be 17, since in this formula the n includes both

maximum and minimum positions.)

5.19 Using $\lambda = \dfrac{\Delta x d}{D}$ then $580 \times 10^{-9} = \dfrac{\Delta x \times 0.2 \times 10^{-3}}{2.5}$

$$\Delta x = 7.25 \, \text{mm}.$$

The distance across four fringes $= 29 \, \text{mm}$.

5.20 *D* is the distance between the slits and screen. *d* is the distance between the two slits. (*a*) *D* is a relatively large quantity so may be measured with some accuracy. Δx the distance between fringes is small but the total distance between a number of fringes may be measured with a travelling microscope and thus Δx determined with a low uncertainty. The readings for Δx may also be repeated over different sets of fringes. *d* is also a small quantity and must be determined directly, this quantity will carry the largest uncertainty of the group into the final determination of λ.
(*b*) If *D* increases, Δx, the distance between fringes, also increases.
(*c*) If the slits are increased too much no interference will be observed as the width of the slits will not be of the same order as the wavelength of the light.
(*d*) The blue fringes are closer together than the yellow fringes.
(*e*) A series of yellow and blue fringes are obtained with the blue ones nearer the central patch.

5.21 (*a*) Microwaves. (*b*) Phosphorescence.

5.22 The time taken for the reflection of the waves will remain at 10^{-5} s as the velocity of the waves remains the same. The 8 cm waves will be reflected with a broader spread, causing a more diffuse signal to be received. Hence shorter wavelength waves enable a more detailed outline of the object to be determined.

Chapter 6

6.1 Heat required $= 10 \times 120 + 0.4 \times 4.2 \times 10^3 \times 10$
$$= 1200 + 16\,800$$
$$= 18\,000\,\text{J}.$$

$$\text{Power} = \frac{\text{heat}}{\text{time}} = \frac{18\,000}{60} = 300\,\text{W}.$$

6.2 Heat supplied by the heater $= 20 \times 2 \times 60\,\text{J}$.
Heat absorbed by the water $= 0.3 \times 4.2 \times 10^3 \times (T - 20)$, where *T* is the final temperature.
Assuming no heat loss,

$$T - 20 = \frac{20 \times 2 \times 60}{0.3 \times 4.2 \times 10^3}$$
$$= 1.9$$

Final temperature $= 21.9\,°\text{C}$.
The final temperature would be lower in a copper container.

6.3 Heat supplied by the heater in one minute
$$= 0.6 \times 4.2 \times 10^3 \times 10$$
$$= 25.2\,\text{kJ}$$
Heat required
$$= \text{heat to raise temperature} + \text{heat to evaporate}$$
$$= 0.03 \times 4.2 \times 10^3 \times 20 + 0.03 \times 2.3 \times 10^6$$
$$= 2.52 \times 10^3 + 69 \times 10^3$$
$$= 71.52\,\text{kJ}.$$
Time taken
$$= \frac{71.52 \times 10^3}{25.2 \times 10^3} \simeq 3\,\text{minutes}.$$

6.4 (a) Pressure = $h\rho g$ $= 0.8 \times 13.6 \times 10^3 \times 10$
$= 1.09 \times 10^5$ Pa.
(b) Pressure = $5 \times 10^3 \times 10 = 5 \times 10^4$ Pa.

6.5 For each atom, the relative atomic mass expressed in g (10^{-3} kg) is the mass of one mole which contains 6×10^{23} atoms.
Thus 6×10^{23} atoms of tin have a mass of 119×10^{-3} kg

$$1 \text{ atom of tin has a mass of } \frac{119 \times 10^{-3}}{6 \times 10^{23}}$$
$$= 1.98 \times 10^{-25} \text{ kg.}$$

Similarly for the other atoms:

Atom	aluminium	platinum	cobalt
Relative atomic mass	27	195	59
Mass of one atom (kg)	4.5×10^{-26}	3.25×10^{-25}	9.8×10^{-26}

Observe the advantage of the relative atomic mass scale when discussing atoms.

6.6 Using pressure = $\dfrac{\text{force}}{\text{area}}$ then $0.2 \times 10^5 = \dfrac{8}{\text{area}}$.
Giving the area of cross section = 4×10^{-4} m^2
and the diameter = 2.3 cm.

When the syringe is loaded, pressure difference = $\dfrac{0.4 \times 10}{4 \times 10^{-4}} = 0.1 \times 10^5$ Pa.

Hence the reading on the gauge is 1.1×10^5 Pa.

6.7 Let x be the volume of the connecting tube
$p = 1.10 \times 10^5$ Pa $V = (40 + x)$ cm^3
$p = 1.35 \times 10^5$ Pa $V = (30 + x)$ cm^3 (a *reduction* of 10 cm^3)
Using $p_1 V_1 = p_2 V_2$ $1.10 \times 10^5 (40 + x) = 1.35 \times 10^5 (30 + x)$
$44 + 1.1x = 40.5 + 1.35x$
$x = 14$ cm.

6.8 (a) Root mean square velocity $c_{\text{r.m.s.}}$

$$= \sqrt{\frac{3p}{\rho}} = \sqrt{\frac{3 \times 10^5}{9 \times 10^{-2}}} = \sqrt{3.3 \times 10^6} = 1.8 \times 10^3 \text{ m s}^{-1} \text{ at s.t.p.}$$

(b) Average kinetic energy of the hydrogen molecule

$= \frac{1}{2}m\overline{c^2}$
$= \frac{1}{2} \times 2 \times 1.67 \times 10^{-27} \times 3.3 \times 10^6$
$= 5.5 \times 10^{-21}$ J at s.t.p.

Assuming the molecular mass of hydrogen is twice the atomic mass.
(c) The average kinetic energy of *all* molecules is the same at the same temperature, $E_k \propto T$. Therefore the average kinetic energy of oxygen is also 5.5×10^{-21} J at s.t.p.

(d) The average kinetic energy varies directly with temperature. $\dfrac{E_{k1}}{E_{k2}} = \dfrac{T_1}{T_2}$.

$E_{k1} = 5.5 \times 10^{-21}$ J, $T_1 = 273$ K, $T_2 = 300$ K,

$E_{k2} = \dfrac{5.5 \times 10^{-21} \times 300}{273} = 6 \times 10^{-21}$ J.

The kinetic energy of the nitrogen molecules at 27°C is 6×10^{-21} J.

6.9   Mean square velocity of hydrogen $= 3.3 \times 10^6$ at $0°C$.

For any gas, $\frac{1}{2}m\overline{c^2} \propto T$.

For hydrogen m remains constant, hence $\overline{c^2} \propto T. \dfrac{\overline{c_1^2}}{\overline{c_2^2}} = \dfrac{T_1}{T_2}$,

so for $\overline{c_1^2} = 3.3 \times 10^6 \text{ m s}^{-1}, \overline{c_2^2} = 10^8 \text{ m s}^{-1}, T_1 = 273 \text{ K}$,

$$T_2 = \frac{273 \times 10^8}{3.3 \times 10^6} = 8.3 \times 10^3 \text{ K}.$$

Note: High temperature is required to increase the velocities of the molecules.

6.10   (a) At constant temperature $m_A \overline{c_A^2} = m_B \overline{c_B^2}$.
The relative molecular or atomic masses are used for m_A and m_B.
For nitrogen $m_B = 28$, and for hydrogen $m_A = 2$

$$2 \times (2.0 \times 10^3)^2 = 28 \times \overline{c_B^2}$$
$$c_{\text{r.m.s.}} = 535 \text{ m s}^{-1} \text{ at } 51°C$$

For neon, $m_B = 20$

$$2 \times (2.0 \times 10^3)^2 = 20 \times \overline{c_B^2}$$
$$c_{\text{r.m.s.}} = 632 \text{ m s}^{-1} \text{ at } 51°C$$

For chlorine $m_B = 71$

$$2 \times (2.0 \times 10^3)^2 = 71 \times \overline{c_B^2}$$
$$c_{\text{r.m.s.}} = 336 \text{ m s}^{-1} \text{ at } 51°C.$$

Chapter 7

7.1   Energy $= hf = \dfrac{hv}{\lambda}$

where v is the velocity of light, h is Planck's constant, and λ is the wavelength

$$\text{Energy} = \frac{6.6 \times 10^{-34} \times 3 \times 10^8}{5.46 \times 10^{-7}}$$
$$= 3.6 \times 10^{-19} \text{ J}.$$

7.2   If the surface of the zinc plate is rusty or has surface impurities, the electrons will require greater energy to escape from the surface so that photoelectric emission will only take place with frequencies *higher* than those usually required. There is also the possibility that the impurities will absorb the radiation.

7.3   Some photoelectrons are nearer the surface or less easily bound and so require less energy to escape than others, which are more strongly bound in the metal interior.

7.4   This depends on the momentum p of the particles. As p increases the associated wavelength decreases. For electrons with their light mass the wavelength is comparable to that of X-rays where the spacing in crystals, used to demonstrate diffraction, is of the same order as these wavelengths. For heavier particles, or objects such as ball bearings, as $p \uparrow$ and $\lambda \downarrow$ there will be no suitable aperture or obstacle with small enough dimensions to give *observable* diffraction effects.

7.5 (a) An absorption is represented by an arrow pointing upwards.
Using $\Delta E = hf$ $(-21.8 - (-87.2)) \times 10^{-19} = 6.63 \times 10^{-34} \times f$
$\qquad\qquad\qquad\qquad\qquad\qquad 65.4 \times 10^{-19} = 6.63 \times 10^{-34} \times f$
Giving $f = 9.86 \times 10^{15}$ Hz (ultra-violet).
(b) For the emission line, using $\Delta E = hf$
$\qquad (-9.7 - (-87.2)) \times 10^{-19} = 6.63 \times 10^{-34} \times f$
$\qquad\qquad\qquad\qquad\qquad f = 1.17 \times 10^{16}$ Hz.
(c) The remaining possible spectral line involves a transition between the upper two levels.
$$\text{Energy difference} = (-9.7 - (-21.8)) \times 10^{-19}$$
$$= 1.21 \times 10^{-18} \text{ J}$$

7.6 (a) Using $E = hf$ $8 \times 10^{-19} = 6.63 \times 10^{-34} \times f$
$\qquad\qquad\qquad\qquad\qquad\qquad f = 1.21 \times 10^{15}$ Hz (ultra-violet).
(b) A frequency lower than this will have insufficient energy. A higher frequency can eject an electron with more kinetic energy.
(c) The velocity depends on E_k. The velocity would *not* be affected by a change of intensity.

7.7 The emission spectrum of an element has coloured lines on a black background. The absorption spectrum has black lines on a continuous spectrum (coloured) background.
   The frequencies of the coloured emission lines correspond to the frequencies of the black absorption lines and are characteristic for each element. The frequencies depend only on the difference in the energy levels.

7.8 In Thomson's model the positive charge was spread throughout the entire atomic volume, hence the electrostatic repulsion on a small α particle would never be sufficient to cause appreciable deflection. With Rutherford's model the concentrated positive charge in the central nucleus exerts a high electrostatic repulsion on the occasional α particle which travels near by, causing the deflections observed, the events at position D being the most significant, Figure 7.14 page 216.

7.9 Isotopes: $^{206}_{82}\text{Pb}$, $^{207}_{82}\text{Pb}$, $^{208}_{82}\text{Pb}$.
Isobars: $^{40}_{19}\text{K}$, $^{40}_{18}\text{Ar}$.

7.10 α and β particles. Neutrons and γ-rays are uncharged.

7.11 Their half lives will be the same.

7.12 $6\frac{1}{4}\%$. After 2 hours, 50% left; after 4 hours, 25% left; after 6 hours, $12\frac{1}{2}\%$ left, and after 8 hours, $6\frac{1}{4}\%$ remains.

7.13 The background count must be subtracted. Thus data to be plotted is:

Time in days	0	1	2	3	4	5
Activity	130	85	56	37	24	15

From the graph the time taken for the count rate to fall from 100 to 50 is
$2.25 - 0.65 = 1.6$ days.

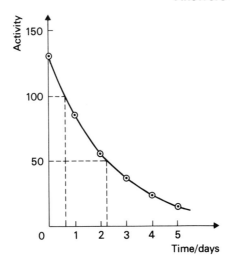

7.14 Momentum is conserved in any collision provided no external forces act on the system.
Energy is conserved in all isolated systems provided the equivalence of mass and energy is taken into account.
Power is not conserved.
Mass is conserved in chemical reactions where no nuclear changes take place.
The total charge in any isolated system is conserved.

7.15 This is an example of nuclear fusion, a larger nucleus, oxygen, is built up from nitrogen. A proton is emitted.

7.16 Mass of $3n + 3p = 10.0425 \times 10^{-27}$ kg.
Mass defect $= 5.59 \times 10^{-29}$ kg.
Binding energy $= 5.031 \times 10^{-12}$ J.

7.17 (*a*) Lead ^{208}Pb will have the larger total binding energy because it has the larger number of nucleons.
(*b*) From the graph, Figure 7.22, nickel has the larger binding energy per nucleon.

7.18 Very high temperatures are required to encourage the positive nuclei to collide. Hence the apparatus must be able to sustain high temperatures. Large magnetic fields are used to accelerate the positive nuclei and to repel them from the sides of a doughnut-shaped container to encourage collisions.

Chapter 8

Individual graphs are not given for all the problems. Students should present their graphs to a teacher for criticism. A glance at a correct graph is usually insufficient to spot all the errors in one's own graph. Attention is drawn to the list of points for graph plotting on page 233.

8.1 (a) Incomplete title, units missing from the x-axis, poor scale on the y-axis which should start from the origin and the last result has not been included. Only a straight line *through* the origin, (0,0) verifies that $y \propto x$. This graph would be a curve if plotted on an improved scale. Boyle's law states that p varies as $\dfrac{1}{V}$.

(b) A graph of p and $\dfrac{1}{V}$ should be plotted, which *does* give a straight line through the origin, (0,0).

8.2 (a) The results suggest a direct relationship, but not $h \propto v$. Consider if $h \propto v^2$ by determining $\dfrac{v^2}{h}$ for the first, third, and last result, these are 0.194, 0.200, and 0.197.

These results suggest that $h \propto v^2$ within the limits of the experiment.
(b) A graph of v^2 against h is plotted, with h on the x-axis, because h is the independent variable. This is a straight line which passes through the origin, (0,0).
(c) Using E_p at the maximum height $= E_k$ at the lowest point
$$mgh = \tfrac{1}{2}mv^2$$

Giving $$g = \dfrac{v^2}{2h}$$

The acceleration due to gravity $= \tfrac{1}{2}$ (gradient of the graph) $\times 100$.
The gradient is about 0.197, giving $g = 9.8\,\mathrm{m\,s^{-1}}$.

8.3 (a) The results suggest an inverse relationship.
Using the first two results, $1.73 \times 10 = 17.3$ and $0.86 \times 20 = 17.2$, which agrees with this choice of an inverse relationship. A graph of energy and $\dfrac{1}{R}$ should be plotted with $\dfrac{1}{R}$ on the x-axis. This is a straight line passing through the origin (0,0), verifying that energy $E \propto \dfrac{1}{R}$.

(b) Energy $= IVt = \dfrac{V^2 t}{R}$

and $V^2 = $ gradient $\times \dfrac{1}{2 \times 60}$ since the gradient is $E/(1/R)$.
The gradient is about 17.3×10^3 giving $V = 12\,\mathrm{V}$. (Remember that the energy is in kJ.)

8.4 (a) The graph, with R on the y-axis, is a straight line which meets the y-axis below the origin at $-2\,\Omega$. Thus $r = 2\,\Omega$.
The gradient gives the e.m.f. ε and is equal to 8 V.

(b) When $R = +2\,\Omega, \dfrac{1}{I} = 0.5$ giving the current $I = 2\,\mathrm{A}$.

8.5 (a) The graph of $\dfrac{1}{v}$ (y-axis) and $\dfrac{1}{u}$ is a straight line with a *negative* slope which cuts the x-axis at $0.055\,\mathrm{cm^{-1}}$ and the y-axis at $0.055\,\mathrm{cm^{-1}}$.

(b) From the formula if $\dfrac{1}{u} = 0$ then $\dfrac{1}{f} = \dfrac{1}{v}$, i.e. the intercept on the y-axis,

and if $\dfrac{1}{v} = 0$ then $\dfrac{1}{f} = \dfrac{1}{u}$, i.e. the intercept on the x-axis.

Hence $\dfrac{1}{f} = 0.055$ and the focal length $f = 18\,\text{cm}$.

8.6 (a)

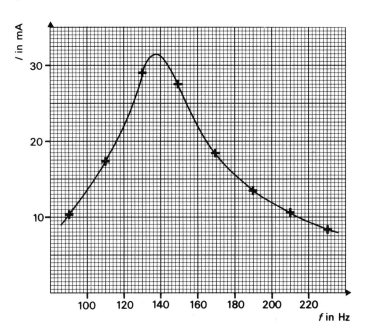

Observe that the frequency scale does not start at zero, but at 80 Hz.

(b) The reactance of both the inductor and capacitor vary with frequency, hence the current will alter.

(c) A smooth curve must be drawn, giving a maximum current of 31.5 mA. Notice that the curve is not symmetrical.

(d) The frequency, for this maximum current, is 138 Hz.

8.7 The absolute uncertainty is 0.1 mA. The percentage uncertainty is 2.9 %.

8.8 $X_c = \dfrac{1}{2 \times \pi \times 110 \times 32 \times 10^{-6}} = 45.2\,\Omega.$

$\%$ uncertainty in $X_c = 9.4 + 1.8 = 11.2\%.$

Actual uncertainty in $X_c = 5.1\,\Omega.$

Giving $X_c = (45.2 \pm 5.1)\,\Omega.$

8.9 $X = \dfrac{b_1 S}{b_2}$

$X = \dfrac{45 \times 110}{55} = 90.0\,\Omega.$

To obtain the $\%$ uncertainty in X add the $\%$ uncertainties in $b_1, b_2,$ and S.

$\%$ uncertainty in $X = 2.2 + 1.8 + 0.9 = 4.9\%.$

Actual uncertainty in X is $4.4\,\Omega.$

Giving $X = (90.0 \pm 4.4)\,\Omega.$

[Alternatively the limiting values could be used.

The maximum value of X is $\dfrac{46 \times 111}{54} = 94.6\,\Omega.$

The minimum value of X is $\dfrac{44 \times 109}{56} = 85.6\,\Omega.$

Giving $X = (90.0 \pm 4.6)\,\Omega.$
The larger uncertainty of 4.6 is chosen.
The answer using the limiting values will tend to differ slightly from that using % uncertainties.]

8.10 (a) Change in temperature = $(11.0 \pm 1)°C$ (uncertainties added).

$$\text{Power} = \frac{mc\,\Delta T}{time} = \frac{0.2 \times 4.18 \times 10^3 \times 11}{120} = 76.6\,\text{W}.$$

% uncertainty in the power value $= 0.5 + 9.1 + 0.2 = 9.8\,\%.$

$$\text{Power} = (76.6 \pm 7.5)\,\text{W}.$$

(b) The temperature, with a % uncertainty of 9.1 %.
(c) If no precautions are taken to prevent heat loss to the container or the room, then the water will *not* receive all the heat supplied by the heater. The temperature of 26°C, is probably too low, and the power of the heater is greater than 76.6 W. The uncertainty is determined from the accuracy of the measuring instruments, *not* from the design of the experiment.

8.11 (a) A graph of t and $\sqrt{h}$ should be plotted.
(b) The intercept on the time axis of 0.04 s suggests a systematic uncertainty in the time measurement. This is probably due to magnetic hysteresis, i.e. the electro-magnet does not lose its magnetism immediately.
(c) From the gradient, g is about $9.7\,\text{ms}^{-2}$.
(d) No, none required.

8.12 $1\,\text{MeV} = 1.602 \times 10^{-13}\,\text{J}$ and $E = mc^2$.

$$\text{Mass of } \mu^- = \frac{105.7 \times 1.602 \times 10^{-13}}{(3 \times 10^8)^2}$$

$$= 1.88 \times 10^{-28}\,\text{kg}.$$

$$\text{Mass of } \tau^- = \frac{1.9 \times 10^3 \times 1.602 \times 10^{-13}}{(3 \times 10^8)^2}$$

$$= 3.4 \times 10^{-27}\,\text{kg}.$$

For interest, observe that the mass of the muon is almost 207 times that of the electron, and the mass of the tau lepton about twice the mass of the proton.

Exercise Section

Mechanics

1 A traveller drives 7 km north, then 7 km due west, then 2 km south-east. What is the distance and displacement from the starting point?

2 Four forces act on an object O as shown.

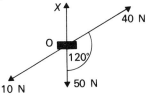

(a) By choosing a suitable value of X is it possible to maintain the object in equilibrium?

(b) What is the resultant force if $X = 20\,\text{N}$?

3 An aeroplane travelling due west at $1200\,\text{m s}^{-1}$ encounters a cross wind of $500\,\text{m s}^{-1}$ blowing due north.

(a) State the resultant velocity.

(b) In which direction should the aeroplane fly in order to remain on a westerly course?

4 A ship steaming due north at $24\,\text{km h}^{-1}$ encounters a cross current of $7\,\text{km h}^{-1}$ due west.

(a) Determine the resultant velocity of the ship.

(b) In what direction and with what velocity must the ship steam in order to effectively travel $24\,\text{km h}^{-1}$ due north?

5 A man is paddling due north in a canoe at $3\,\text{m s}^{-1}$ up a river when he meets a strong wind blowing at $1.25\,\text{m s}^{-1}$ due west. What is the relative velocity of the canoe to the bank? If the canoe was originally in the middle of the river, which is 10 m wide, how long did it take before the canoe hit the bank, assuming the strength of the wind remained constant?

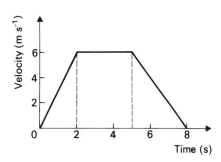

6 This graph of velocity against time shows the motion of a boy running in a straight line, away from a point P. Determine:

(a) the acceleration in the first two seconds,

(b) the total distance travelled,

(c) the average velocity for these 8 seconds,

(d) Calculate the displacement at the end of each second and plot a displacement/time graph for these 8 s, giving numerical values on both axes.

7 An object was thrown vertically upwards with a velocity of $35\,\text{m s}^{-1}$. What was its position after 6 s?

8 An object is released from an aircraft travelling upwards at $25\,\text{m s}^{-1}$.

(a) Sketch speed/time and velocity/time graphs for the first 4 s.
(b) Sketch a graph of kinetic energy E_k with time for these 4 s. Numerical values are required on the time axes.

9 A boy runs 300 m to a letter box. He takes 75 s to reach the box, pauses for 3 s to post the letter, and then returns in $1\frac{1}{2}$ minutes.

(a) Draw a displacement/time graph.
(b) Sketch a velocity/time graph for the complete trip, giving numerical values on both axes. State any assumptions made.

10 When an object moves along a horizontal surface, the following velocity/time graph is obtained.

(a) Did the time measurements commence when the object started to move?
(b) What is the acceleration in the first 10 s, between the 40th and 45th s, and in the last 10 s?

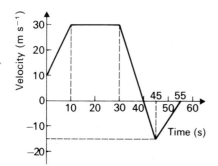

(c) Sketch a displacement/time graph. Numerical values are required only on the time axis, but it should be clear whether the graph is a straight line or a curve.
(d) Sketch an acceleration/time graph giving numerical values on both axes.

11 An object has an acceleration of 2 m s^{-2}. Complete the table below showing the instantaneous velocities and displacements at the *end* of each second, and the average velocity *over* each successive second.

time (s)	1	2	3	4	5	6
instantaneous velocity (m s^{-1})	2	4				
displacement (m)	1	4				
average velocity (m s^{-1})	1					

From the table state: (a) the average velocity over the 4th s, (b) the velocity after 4.5 s, (c) the distance travelled after 5 s.

12 A ball is thrown vertically upwards with a speed of $55\,\text{m s}^{-1}$. Determine its distance and displacement after 7 s.

13 A space rocket of total mass 4×10^5 kg, is travelling in outer space. When the motor is fired, it exerts a constant force of 7×10^6 N, and fuel is used up at the rate of $2 \times 10^3\,\text{kg s}^{-1}$. The motor is fired for one minute then switched off.

(a) Determine the acceleration at the beginning and end of this one-minute period and sketch force/time and acceleration/time graphs for the two minutes commencing when the motor is switched on. Numerical values are required on the axes.
(b) Sketch a velocity/time graph for these same two minutes with numerical values on the time axis only.
(c) Using the acceleration/time graph, state the change in velocity of the rocket due to this firing.

14 An object is knocked off the end of a table with a horizontal speed of $20\,\text{m s}^{-1}$. If the vertical height of the top of the table above the floor is 1.25 m determine:

(a) the time taken for the object to hit the floor,
(b) the resultant velocity just before impact (magnitude and direction).

15 An object is projected at an angle of 30° to the horizontal with a velocity of $24\,\mathrm{m\,s}^{-1}$. Neglecting air resistance, calculate:

(a) its position and velocity in the horizontal and vertical directions after one second,

(b) the greatest height reached,

(c) the total distance travelled and the time for it to return to the ground.

(d) Draw *separate* velocity/time graphs for the horizontal and vertical velocities for the entire flight of the object. Numerical values are required on the axes.

16 A pellet leaves an air gun with a velocity of $100\,\mathrm{m\,s}^{-1}$ at an angle of 30° to the horizontal. Calculate:

(a) the maximum height reached,

(b) the total time of flight,

(c) the horizontal distance travelled.

17 A boy points his arrow horizontally at the central spot of a target. If the arrow leaves his bow with a velocity of $50\,\mathrm{m\,s}^{-1}$ and strikes the target 20 cm below the spot, how far from the target was he standing?

18 The diagram illustrates a multiple flash photograph of a moving ball. The flash rate is 15 per second.

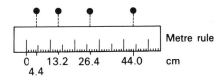

(a) Calculate the acceleration of the ball.

(b) Why was the metre rule included?

(c) What do you think was happening to the ball?

(d) Explain what difference, if any, you would expect in the result to part (a) if a ticker timer was used to determine the acceleration.

19 One ball A was dropped freely at the same instant that another ball B was thrown horizontally. The diagram of the multiple flash photograph indicates the subsequent motion of the two balls. The distances marked are the actual distances traversed.

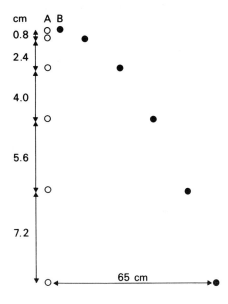

(a) Calculate the flash rate.

(b) Find the velocity with which B was thrown horizontally.

(c) Give one advantage of multiple flash methods over ticker timer methods.

(d) What else might you expect to see included in the photograph?

20 A velocity can be measured using (a) a ticker tape, (b) a light gate, or (c) strobe photography. Describe each method, stating clearly what measurements are required in order to calculate the velocity. List the advantages and disadvantages of each method.

21 A cannon is situated at the top of a 45 m cliff. A ship is moving towards the cliff at $8\,\mathrm{m\,s}^{-1}$. The cannon is fired horizontally when the ship is 102 m away.

(a) Assuming that the ship is hit, calculate the speed of the cannon ball as it leaves the cannon.

(b) Calculate the velocity of the cannon ball when it hits the ship.

22

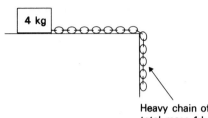

Heavy chain of total mass 1 kg

The above system is released from rest when half the chain is over the side of the bench. Neglecting friction, describe the acceleration of the 4 kg mass.

23 A conker is whirled in a horizontal circle with a uniform speed of $0.5\,\mathrm{m\,s^{-1}}$. If the circumference of the circle is 2 m:

(a) how long does it take the conker to make one revolution?
(b) What change in velocity occurs in 1 s?
(c) What would happen to the conker if the string was cut?

24 Refer to the data booklet and determine the weight of a 5 kg block on the surface of (a) the Earth, (b) the Moon, (c) Mars, (d) Venus.

25 On a certain planet an object falls vertically 9 m, from rest, in 3 s.

(a) Determine the gravitational field strength on this planet.
(b) A stone is thrown horizontally from a cliff at $8\,\mathrm{m\,s^{-1}}$ and hits the ground 4 s later. State the height of the cliff, and the speed with which the stone hits the ground, on this same planet.

26 A trolley of mass 1.5 kg, initially at rest, is pulled along a horizontal board with a force of 5 N. If the frictional resistance is 0.5 N:

(a) what is the velocity of the trolley after 1 m?
(b) How would you reproduce this motion experimentally?
(c) Explain in detail how you would determine the velocity at this 1 m position.

(d) How could you reduce considerably or eliminate the frictional resistance?

27 A trolley of mass 2 kg is at rest on a horizontal floor. A force of 4 N acts for 3 s. If the frictional resistance is 1 N throughout, describe the position and velocity of the object after 6 s.

28 A trolley of mass 1.2 kg is situated on a plane. Determine the component of the weight down the plane, and hence the acceleration, if the angle of inclination of the plane to the horizontal is:

(a) 45°,
(b) 30°.
(c) How would these answers be affected if there is a frictional resistance of 1 N?

29 A puck slides down a rough plane. The following table indicates the distance travelled at the end of each second after the object is released.

time	0	1	2	3	4
distance	0	0.2	0.8	1.8	3.2

(a) Calculate the acceleration of the puck.
(b) State the average velocity in the third second.
(c) Estimate the velocity at the end of the fifth second.
(d) If the plane is inclined at an angle of 11.5° to the horizontal, what acceleration would you expect if there is no frictional force?
(e) If the mass of the puck is 500 g, calculate the frictional force.

30 An object of mass 2 kg is projected with a velocity of $4\,\mathrm{m\,s^{-1}}$ up a plane inclined at 30° to the horizontal. How far up the plane does it go if:

(a) there is no frictional resistance?
(b) the frictional resistance is 2 N and independent of the speed?

31 What is the least force which must be supplied to a helicopter of mass $2 \times 10^3\,\mathrm{kg}$ to enable it to lift off vertically with an initial acceleration of $5\,\mathrm{m\,s^{-2}}$, if the air resistance is $2\,\mathrm{N\,kg^{-1}}$?

32 A piece of cord is attached to a trolley of mass 0.9 kg held at one end of a plane. The cord is then passed over a pulley situated at the other end of the plane and a 100 g mass is attached and allowed to hang vertically. The trolley is released. Determine the acceleration of the trolley and mass system if:

(a) the plane is friction compensated,
(b) there is a 0.2 N frictional resistance.

33 A cyclist of mass 48 kg has to overcome air resistance of 20 N when travelling along a level road at $6 \, \text{m s}^{-1}$. Calculate:

(a) the average force he exerts,
(b) the power he must supply,
(c) his velocity if he ascends a hill of 1 in 8 (measured along the slope), assuming his power output remains unchanged.

34 A truck of mass 10^3 kg with a power output of 7.5 kW ascends a hill of 1 in 20 (measured along the slope). If the acceleration is $0.2 \, \text{m s}^{-2}$ when the car is moving at $9 \, \text{m s}^{-1}$, determine:

(a) the resultant force on the truck,
(b) the force exerted by the engine,
(c) the frictional resistance.

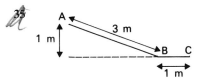

A ball of mass 1.5 kg is placed at the top of a rough plane AB where the frictional force is 1 N. BC is a smooth horizontal surface. Determine:

(a) the potential energy at A,
(b) the work done against friction when travelling from A to B,
(c) the kinetic energy at B,
(d) the kinetic energy at C,
(e) the velocity at C.
(f) Explain carefully how you would measure experimentally the velocity just before C.

36 A bullet is fired from a rifle into a block of plasticine attached to a stationary trolley on a rough horizontal plane. The bullet passes through the plasticine. Describe the motion of the trolley and the bullet. Discuss in detail the energy changes taking place from the moment the bullet is fired until both the bullet and trolley are at rest.

37

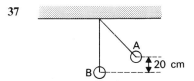

A pendulum bob is pulled to one side such that it is 20 cm vertically above its rest position. The mass of the bob is 30 g.

(a) Determine the change in potential energy taking place when the bob falls from A to B.
(b) What is the velocity of the bob at B?
(c) How could this velocity at B be measured experimentally?
(d) State the acceleration of the bob at B.
(e) If the bob takes 0.1 s to fall from A to B what is the period of the pendulum and the frequency of the oscillations?

38 An 80 kg man stands on scales placed on the floor of a lift. Describe the *two* possibile motions of the lift in each case, if the scales read: (a) 800 N, (b) 720 N, (c) 960 N.

39 A 30 g box of side 5 cm is given a sharp tap causing it to slide along a smooth horizontal table. It interrupts a light beam directed on to a photocell. An electric clock, connected to the photocell, records a time of 0.025 s. The box then falls off the edge of the table. If the height of the table is 0.8 m, determine:

(a) the momentum with which it leaves the table,
(b) the time taken to reach the ground,
(c) the precise position where the object lands,
(d) the horizontal and vertical components of the velocity just before impact with the ground,
(e) the kinetic energy just before impact.

40   A ball of mass 4 kg accelerates from rest to $30\,\mathrm{m\,s^{-1}}$ in 5 s. Sketch graphs of momentum against velocity, and kinetic energy against velocity for these 5 s.

41   Two marbles A and B each have a mass of 40 g. A rolls down a smooth chute, strikes B which is stationary at the bottom, and drops vertically downwards. B hits the ground at P.
Determine:

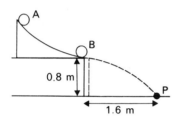

(a)   the velocity of B as it leaves the bottom of the chute,
(b)   the momentum of B after the collision,
(c)   the kinetic energy of A just before it lands on the ground,
(d)   the kinetic energy of B just before it lands at P,
(e)   the momentum of B just before it lands at P.
(f)   How is the position P determined experimentally?

42   A ball of mass 50 g strikes a wall at an angle of 45° and rebounds with the same speed as shown.
Find:

(a)   the change of momentum of the ball,
(b)   the impulse of the ball on the wall,
(c)   the impulse of the wall on the ball,
(d)   the force of impact of the ball on the wall if the impact lasts 0.3 ms.

43

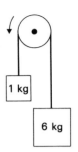

A motor rated at 200 W is used to turn the pulley wheel, of circumference 0.3 m, illustrated above. If the weights remain stationary when the wheel makes 10 revolutions per second, determine the efficiency of the motor.

44   A 100 g rubber ball at rest, is hit with a bat. The bat is in contact with the ball for 2.4 ms, and the ball moves off at $6\,\mathrm{m\,s^{-1}}$. Draw a force/time graph with numerical values on the axes to illustrate this impact.

45   A trolley of mass 1 kg, travelling on a smooth horizontal plane with a speed of $4\,\mathrm{m\,s^{-1}}$, collides with another stationary trolley of mass 2 kg and rebounds back along its initial path with a speed of $1.4\,\mathrm{m\,s^{-1}}$.

(a)   Determine the velocity acquired by the 2 kg trolley.
(b)   Describe how this velocity could be measured experimentally using a ticker tape. Comment on any disadvantages.

46   An object of mass 100 g is dropped from a height of 2 m and rebounds to a height of 1.8 m. Determine:

(a)   the kinetic energy before and after impact with the ground,
(b)   the change of momentum at the ground,
(c)   the impulse on the ground,
(d)   the force of impact of the ball on the ground if the impact lasts 3 ms.

47   In an experiment to determine the force of impact when kicking a ball, the time of impact was measured on an electric clock and found to be 22 ms. The pupil then recorded other results in her book as follows:

mass of ball $= 3.3\,\text{kg}$,
diameter of the ball $= 20\,\text{cm}$.

After a calculation she stated that the force of impact was $1.5 \times 10^3\,\text{N}$. Explain in general terms how she calculated this force. What measurement did she fail to record in her book and what was its value?

48 A $10\,\text{kg}$ metal box falls for $0.6\,\text{s}$ vertically downwards and hits a short $2\,\text{kg}$ pole $20\,\text{cm}$ into the ground.

(a) At what speed do the box and pole move into the ground.
(b) Calculate the average resistive force of the ground to the pole and box.

Electricity

49 A charge of $0.3\,\text{mC}$ is placed midway between two plates separated by $10\,\text{cm}$. The electric field strength is $200\,\text{V m}^{-1}$.

(a) Explain what is meant by electric field strength.
(b) Calculate the p.d. between the plates and the force on this charge.

50 The gun p.d. of a c.r.o. is $1200\,\text{V}$. The Y plates are $2\,\text{cm}$ long and $0.5\,\text{cm}$ apart. A screen is placed $30\,\text{cm}$ beyond the plates.

(a) Describe the motion of the electron from the moment it leaves the gun until it strikes the screen.
(b) Calculate the velocity with which the electron enters the Y plates.
(c) Calculate the acceleration of the electron between the Y plates when a p.d. of $2\,\text{V}$ is applied to these plates. (Assume that the electron enters the Y plates in a direction parallel to the plates.)
(d) Determine the vertical component of the velocity of the electron on leaving the plates.
(e) State the angle at which the electron is travelling as it leaves the Y plates.
(f) State the deflection produced on the screen to the nearest mm. Hence comment on the need for a Y-gain control on the c.r.o.

51 Explain, with the aid of diagrams, what is meant by a radial electric field and a uniform electric field.

52 In a demonstration Millikan experiment, a small polystyrene ball of mass $10^{-5}\,\text{kg}$ remains stationary between two parallel plates $0.1\,\text{m}$ apart. The electric field strength between the plates is $20 \times 10^3\,\text{N C}^{-1}$.
Find:

(a) the potential difference between the plates,
(b) the charge on the ball in coulombs. State any assumptions involved,
(c) the number of electronic charges on the ball.

53 In a Millikan type experiment, the p.d. (V) between the plates a distance (d) apart is adjusted until a small drop of oil remains stationary.

(a) How could the charge on the drop be altered?
(b) Neglecting the upthrust of the air, state the two forces acting upon the drop.
(c) A p.d. of $2.2\,\text{kV}$ is required to maintain the drop at rest. If the distance between the plates is $6.4\,\text{mm}$ and the mass of the drop is $3.3 \times 10^{-14}\,\text{kg}$, estimate the number of electrons on the drop.

54 In a simple cathode ray tube a heated cathode emits electrons which are attracted to a cylindrical anode. After passing through the anode, the electrons impinge on a fluorescent screen $20\,\text{cm}$ beyond the anode. The p.d. between cathode and anode is $4\,\text{kV}$, and the anode/cathode separation is $10\,\text{cm}$. Find:

(a) the force on an electron, (i) on leaving the cathode, (ii) just before arrival at the anode,
(b) the energy acquired by an electron on reaching the anode, if it leaves the cathode with zero velocity,
(c) the velocity of the electron at the anode,

(d) the velocity of the electron just before striking the screen,
(e) the power output if 10^{16} electrons strike the screen in one second.

55 Explain in general terms how the unit, the ampere, is defined. If a current of one ampere flows down a wire how many electrons pass a point P in 2 s?

56 The resistances of three 1 m lengths of wire were found to be $1\,\Omega$, $4\,\Omega$, and $0.25\,\Omega$. What was the ratio of their diameters, assuming they were made of the same material?

57

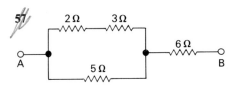

Using the circuit shown, determine:

(a) the total resistance AB,
(b) the currents through the $6\,\Omega$ and $5\,\Omega$ resistors, if the current through the $2\,\Omega$ resistor is 2 A,
(c) the p.d. AB.

58 Three $20\,\Omega$ resistors are available. State the possible resistance values which can be obtained by combining *all* three resistors in different ways.

59 When determining the resistance of a bulb the following results were obtained:

potential difference (V)	1	1.5	2	2.5	3
current (A)		0.18	0.25	0.29	0.31	0.36

(a) Calculate the resistance of the bulb for each set of results.
(b) Comment on any departure from Ohm's law.

60

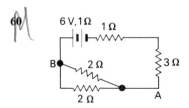

In the figure, the 6 V battery has an internal resistance of $1\,\Omega$.
Calculate:

(a) the current through each of the $2\,\Omega$ resistors,
(b) the p.d. across each $2\,\Omega$ resistor,
(c) the potential of B if A is earthed,
(d) how much energy is dissipated in the $3\,\Omega$ resistor, in 1 s.

61 In the circuit shown, each battery has an internal resistance of $1\,\Omega$.

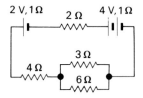

Calculate:

(a) the current through the $2\,\Omega$ resistor,
(b) the p.d. across the $4\,\Omega$ and $6\,\Omega$ resistors,
(c) the p.d. across the $4\,\Omega$ resistor if the 2 V battery is reversed.

62 Using the circuit, determine:

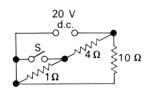

(a) the p.d. across the $10\,\Omega$ resistor when (i) S is open, (ii) S is closed,
(b) the p.d. across the $4\,\Omega$ resistor when (i) S is open, (ii) S is closed,
(c) the current through each resistor when (i) S is open, (ii) S is closed.

63

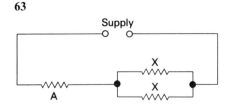

Refer to the circuit shown. The energy converted per second in resistor X is 10 W. Determine the power dissipated in the resistor A if the resistance of A is three times the resistance of X.

64

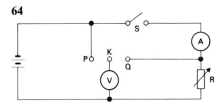

In the above circuit R has a range of 1 Ω–99 Ω. With S open, P is joined to K and the voltmeter reads 8 V. With S closed, and Q joined to K, the voltmeter reads 6.8 V and the ammeter 0.4 A. State:

(a) the e.m.f. and internal resistance of the source,
(b) the value of R at this setting,
(c) the power supplied, and the power transferred to R,
(d) how the power transferred to the resistor could be increased and state its maximum value,
(e) how the p.d. across the resistor R could be increased and state this maximum value.

65 A centre tapped variable resistor of length 48 cm and resistance 24 Ω is connected across a 20 V supply of internal resistance 1 Ω. Calculate:

(a) the p.d. per ohm across the resistor,
(b) the potential gradient of the resistor in V cm^{-1},
(c) the length of resistor required to provide a p.d. of 14 V to an external circuit.

66 A moving coil galvanometer has a resistance of 50 Ω and a f.s.d. of 1 mA. How could you adapt this galvanometer as:

(a) an ammeter with a f.s.d. of 0.5 A,
(b) as a 0–5 V voltmeter,
(c) if the galvanometer has a scale of 50 divisions find the sensitivity per division of the ammeter in (a) and the voltmeter in (b).

67 A sensitive galvanometer is marked 40 Ω, 5 mA.

(a) How could you adapt this galvanometer as a voltmeter to measure p.d. up to 5 V?

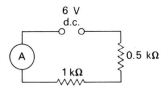

(b) From the circuit estimate the p.d. across the 1 kΩ resistor, assuming the resistance of the ammeter is about 1 Ω.
(c) What would the voltmeter in (a) read if placed across the 1 kΩ resistor?
(d) Comment on the difference in the answers to (b) and (c).
(e) Why did the reading on the ammeter rise when the voltmeter was included in the circuit?
(f) Would the voltmeter give a more 'accurate' measurement of the p.d. across the 0.5 kΩ resistor?
(g) What would happen to the reading on the *ammeter* if a 5 Ω resistor were connected across it?

68

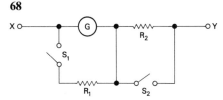

A sensitive galvanometer Ⓖ, shown in the figure, has a resistance of 500 Ω and gives a maximum deflection when 0.1 mA flows through it. State the value of R$_1$ and R$_2$ and the setting of the switches S$_1$ and S$_2$ in order to measure:

(a) a maximum current of 1 A entering at X,
(b) a p.d. of 5 V between X and Y.

69 Describe how you would protect a sensitive galvanometer used in a Wheatstone bridge circuit. Why must the galvanometer be sensitive?

70 Draw a circuit diagram of a *metre bridge* that would enable the resistance of a resistor in the range 4.7 Ω to 4.8 Ω to be measured. Standard resistors of values 4 Ω, 5 Ω, and 6 Ω are available.

(a) Explain the experimental procedure for determining the balance point and indicate how the unknown resistance is calculated.

(b) Which standard resistor is used and why?

(c) Comment on the accuracy, sensitivity and range of the galvanometer used.

71 A student is trying to connect a Wheatstone bridge circuit to verify the balance condition equation. He has five resistors available (1200 Ω, 600 Ω, 400 Ω, 300 Ω, and 100 Ω) a sensitive galvanometer, a battery, and a switch. Draw and label a circuit diagram of the Wheatstone bridge using these components. Explain the use of the fifth resistor.

72 A metre bridge at its balance point has a 30 Ω resistor in the left hand arm and a 50 Ω resistor in the right arm.

(a) Determine the distance along the wire, from the left hand end, of the balance point.

(b) Without altering the balance point, the 30 Ω resistor R is increased to 32 Ω and a current of 1 mA is recorded by the galvanometer. What happens to the galvanometer reading when (i) R is increased to 36 Ω and (b) R is reduced to 27 Ω?

73

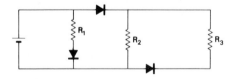

The resistors R_1, R_2, and R_3 have the same resistances. Compare the values of the current through each resistor.

74

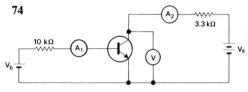

(a) Compare the readings on the ammeters A_1 and A_2.

(b) What will happen to the reading on Ⓥ if the battery V_s is reversed.

(c) How will the reading on A_1 and A_2 alter, if at all, if the battery V_b is reversed?

(d) What type of transistor is shown in this circuit?

75 A supply is labelled $V_{r.m.s.} = 120$ V. Explain what is meant by the term $V_{r.m.s.}$. This supply is connected to 20 Ω and 40 Ω resistors in series. Sketch the trace which would be observed on a c.r.o. placed across the 20 Ω resistor. Give a numerical value.

76 A d.c. supply of value 4 V and an a.c. supply of $V_{r.m.s.} = 4$ V, are connected in turn across a 2 kΩ resistor. Sketch on the *same* axes the traces observed on a c.r.o. connected across the 2 kΩ resistor.

77 In an experiment to demonstrate the effective voltage of an a.c. source as described in Experiment 3.3 (page 93), explain why it is important that the lamp has the same brightness when the switch is at S_1 or S_2.

78 An a.c. supply is labelled 30 V, 50 Hz. Draw a graph showing the variation of p.d. with time, giving numerical values on the axes. State the peak voltage and peak-to-peak voltage of this source.

79

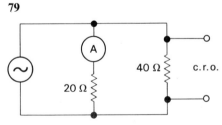

The ammeter records a current of 0.1 A. State the height of the trace in cm if the c.r.o. time base is set at 500 mV cm^{-1}.

80 Given a moving coil galvanometer and four diodes, draw a circuit diagram to show how you could make an a.c. galvanometer. Would you prefer p-n junction diodes or thermionic diodes?

81 Describe a simple experiment to demonstrate that a capacitor can store charge. Refer to this experiment, and explain why work must be done to charge the capacitor.

82 A battery of e.m.f. 6 V is used to charge a 250 μF capacitor. Determine the charge on the plates and the energy that must have been provided by the battery to charge the capacitor.

83 A charged capacitor is connected to an electroscope, as shown in the diagram.

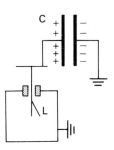

Describe the effect on leaf L when another uncharged capacitor is connected:

(a) in series,
(b) in parallel with the capacitor C.

84 A charged parallel plate capacitor is connected across a gold leaf electroscope. Comment on the divergence of the leaf if:

(a) the plates are moved further apart,
(b) a block of perspex is placed between the plates,
(c) the charge on the plates is increased.

85 In the circuit shown, the switch K vibrates between X and Y at a rate of 50 Hz

when the reading on the milliammeter is 75 mA.

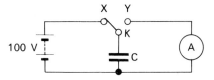

(a) Calculate the charge passing through the milliammeter each time K touches Y.
(b) Find the capacitance of C.
(c) Determine the reading on the ammeter if: (i) another identical capacitor is connected in parallel with C, (ii) the switching frequency is doubled.

86 A parallel plate capacitor has a dielectric constant of 7 and a working voltage of 100 V.

(a) Explain what is meant by the terms: dielectric, dielectric constant, and working voltage.
(b) A student remarks that if a dielectric constant is increased the working voltage can also be increased. Explain why this is not necessarily true.

87

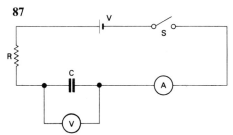

The switch S is closed.

(a) Sketch graphs of the ammeter reading against time, and voltmeter reading against time after S is closed.
(b) Calculate the maximum reading on the ammeter and the maximum reading on the voltmeter if R = 100 kΩ, V = 20 V, C = 32 μF.
(c) Explain clearly how the graphs in (a) would alter if either (i) R was increased to 200 kΩ or (ii) C was reduced to 16 μF.

88 A generating station produces electricity at the rate of 1000 kW and at 250 V a.c. It is delivered to the nearest town through transmission lines whose total resistance is $\frac{1}{20}\,\Omega$. Find the power loss in the lines when:

(a) the electricity is supplied directly to the lines,
(b) a 1:50 step up transformer is used before the transmission lines. State any assumptions that have been made.

89 The secondary of a 40:1 step down transformer is connected to three bulbs rated at 0.75 W, 2.5 V.

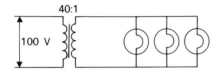

(a) By considering the turns ratio, deduce the p.d. across the secondary circuit.
(b) Determine the secondary current if the bulbs are operating at their rated value.
(c) If the primary current is 0.025 A, find the apparent efficiency of the transformer.
(d) Give *two* reasons why the transformer is not 100% efficient.
(e) A pupil states that the value of the primary current is affected by the value of the secondary current. By determining the value of the primary current when one bulb is removed from the secondary circuit, find if this is correct. Assume that the remaining bulbs operate at their rated value and that the efficiency remains unchanged.

90 The p.d. across the primary has an r.m.s. value of 10 V. The turns ratio of the transformer is 1:3.

(a) Assuming no transformer losses, sketch the trace, giving a numerical value, observed on a c.r.o. placed across R.

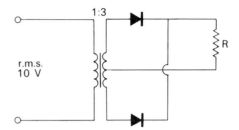

(b) Explain the trace observed when a 20 V d.c. source is used in place of the 10 V a.c. supply.

91 Electricity is transmitted along cables from a generating station to a town 40 km away. The cable consists of aluminium wires of total resistance 6 Ω and steel wires of total resistance 100 Ω. The transmission system operates at 132 kV.

(a) State with a reason if it is desirable to have transmission lines with a low or a high resistance.
(b) Calculate the overall resistance per kilometre of the cable.
(c) Give a reason for the use of both the aluminium and steel wires.
(d) State the turns ratio of the transformer needed at the town to produce a standard 240 V supply.
(e) Explain why a d.c. supply is not used to transmit electricity.

92 The inductor L in the circuit shown has a large number of turns.

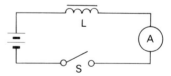

Sketch graphs on the same axes to show how the current varies with time when S is closed, and:

(a) the inductor has a core in position,
(b) the core is removed.

93 Explain, with an example, why Lenz's Law is a result of the law of conservation of energy.

In a motor, a coil carrying a current rotates in a magnetic field, Explain, using Lenz's Law, why a back e.m.f. ε_B is induced in the coil, and hence discuss the equation $V - \varepsilon_B = IR$, where V is the p.d. applied to the coil of resistance R, and I is the working value of the current. Comment on the magnitude of the *initial* current when the motor is first switched on.

94 Explain how an e.m.f. can be induced across a coil through which a current is flowing. Comment on the direction of this induced e.m.f. How can the direction of the induced e.m.f. be reversed?

95

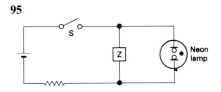

(a) The contact of the switch S vibrates at 2 Hz and the lamp is observed to flash on and off. What is component Z? Explain why the lamp flashes.
(b) The component Z is replaced by another component Y, the vibrator disconnected and the switch S closed. Again the lamp flashes on and off. What is component Y? Again explain why the lamp flashes.
(c) For each of component Z and Y state how the flash rate of the lamp could be increased.

96 Using a constant voltage supply in a series circuit containing a component Z and a milliammeter, the frequency of the supply is gradually increased.

(a) Draw a graph to show the variation of current with frequency when (i) Z is an inductor, and (ii) Z is a capacitor.
(b) What information on the variation of the reactance of a capacitor or an inductor with frequency can be drawn from these graphs?
(c) Both resistance and reactance are a measure of the opposition to current flow. What is the chief difference between them?

97 In the following circuit, at a certain frequency of the a.c. supply, the high resistance voltmeter reads 1.5 V.

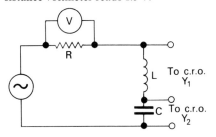

(a) Sketch, with an appropriate numerical value, the trace observed on a c.r.o. placed across the resistor R.
(b) Comment on the values of the current flowing through each component.
(c) The Y terminals of a double beam oscilloscope are connected across the inductor and capacitor as illustrated. Sketch the traces observed if the frequency of the supply is: (i) 5 Hz, (ii) 10 kHz.
(d) What physical quantity does the c.r.o. indicate in part (c).
(e) State the important advantage of a c.r.o. over a voltmeter.

98 In the circuit shown, the e.m.f. of the supply is kept constant. The frequency of the source is steadily increased.

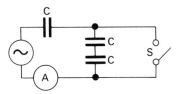

Sketch graphs, on the same axes of the current recorded on the ammeter against frequency when: (a) S is open, (b) S is closed.

99 In the circuit shown, the e.m.f. of the supply is kept constant. The frequency of the source is steadily increased.

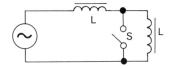

Sketch graphs, on the same axes, of the current recorded on the ammeter against frequency when: (a) S is open, (b) S is closed.

100

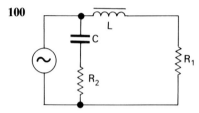

In the circuit shown, R_1 and R_2 are both $1\,k\Omega$ resistors. Which component, R_1, R_2, C, or L, dissipates the most energy when: (a) the frequency is high?, (b) the frequency is low?

101 An inductor is sometimes called a choke. A capacitor can be referred to as a blocking capacitor. What does the inductor 'choke' or the capacitor 'block'? Illustrate your answer with circuit diagrams.

102 The output from a full wave rectifier is connected to a capacitor and an inductor as illustrated.

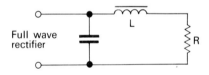

(a) Sketch the trace observed on a c.r.o. placed across R.
(b) Consider the capacitor and inductor separately and explain the effect these components have on the varying unidirectional input.
(c) Would these components be more or less effective at higher frequencies? Why?

Optics and radiation

103 The diagram shows a ray of light incident on a perspex block immersed in water. Determine the angle of refraction in the perspex.

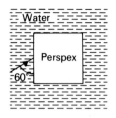

104 Light from an underwater source is refracted out into air. Determine the angle of refraction if the angle of incidence in the water is (a) 35°, (b) 45°, (c) 55°. Illustrate your answer with a sketch showing all three rays.

105 A 45° isosceles glass prism is used to change the direction of light as illustrated in Figure 5.8(a), page 142. What is the minimum value of refractive index possible for this glass?

106

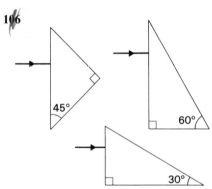

The prisms shown above are made of glass of refractive index 1.7. Calculate the critical angle for this glass. Indicate the path of the rays through the prisms. No angles need be calculated but the direction in which the light is bent should be clearly indicated.

107 Explain the importance of the critical angle in fibre optics.

108
(a) Draw a ray diagram to illustrate the nature of the image of an object placed 20 cm from a lens of focal length 8 cm. Determine the magnification.

(b) State how this image will change as the object is moved towards the lens until it is only 12 cm from the lens.

(c) Draw another ray diagram with the object at this 12 cm position and determine the magnification.

(d) Comment on the change of magnification over this range of object distances.

109 An object is positioned 25 cm from a converging lens. Find, by ray diagrams or otherwise, the nature of the image and the magnification if the focal length of the lens is: (a) 15 cm, (b) 22 cm.

110 A slide projector has a lens of focal length 10 cm. The distance of the slide from the lens is 10.5 cm. A clear image is obtained on a screen 2.1 m away.

(a) If the slide is 35 mm wide, what is the width of the picture?

(b) The screen is 1 m wide. What must be done to make the picture fill the screen?

(c) This enlarged picture is not in focus. How may it be refocused?

111

(a) Draw ray diagrams to illustrate the nature of the *two* different magnified images which may be produced by a converging lens.

(b) State the difference between a converging lens and a diverging lens.

112 Distinguish between deviation and dispersion. Illustrate your answer.

113 A narrow beam of red light from a helium-neon laser enters a glass prism with an angle of incidence of 55°. The angle of refraction is observed to be 31°. Calculate the wavelength, frequency and speed of the light inside the glass. (Relevant data can be found in the science data booklet.)

114 The velocity of sound is increased fourfold on travelling from water into granite. Estimate the wavelength of sound in the water if the wavelength of the sound in the rock is 0.5 m. Draw a sketch to indicate how the direction of the wave alters if it emerges obliquely at the rock surface.

115 In an experiment to study the variation of illuminance E with distance d from a source of light, a set of results are obtained and a graph of E and $\frac{1}{d}$ plotted.

(a) What does illuminance measure?

(b) Was the graph of E and $\frac{1}{d}$ a straight line? Give a reason for your answer.

(c) If the illuminance at 35 cm is 45 units, what is the illuminance at 25 cm?

116 Speed is defined as the distance travelled in one second. Show that the wave equation $v = \lambda f$ for continuous waves is compatible with this definition of speed.

117 In a ripple tank of length 0.8 m, a bar produces a wave every 0.2 s. The distance between consecutive crests is 2 cm.

(a) How long will one crest take to travel the length of the tank?

(b) After a while, standing waves are observed in the tank. Explain how these are formed and state the distance between two maxima (anti-nodes).

118 Plane waves of wavelength 2.5 cm are generated in a ripple tank. The waves are incident normally on a metal barrier containing two slits, each of width 2 cm. Sketch and explain the pattern observed beyond the slits. Outline an experiment using the *same* barrier which demonstrates the wave nature of electromagnetic radiation.

119 Red and blue light both have a speed in air of $3 \times 10^8 \, \text{m s}^{-1}$.

(a) Compare the wavelengths in air of red and blue light.

(b) Do red and blue both undergo the same *reduction* in wavelength when they pass into glass? Give a reason for your answer.

120 Explain with the aid of a diagram, what is meant by constructive and destructive interference of microwaves.

121

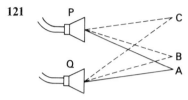

PA = PB = PC = 25 cm
QA = 25 cm QB = 30 cm

Two loudspeakers P and Q are connected to the same signal generator as shown. A microphone, connected to a c.r.o. is used to detect positions of maximum and minimum. A maximum is detected at A, and minima at B and then C.

If the velocity of sound in air is 330 m s^{-1}, determine:

(a) the wavelength of the sound waves,
(b) the frequency employed,
(c) the distance QC.

122 Two narrow lines are scratched on a blackened microscope slide. Describe and explain in terms of the wave nature of light what is observed when a distant tungsten lamp is viewed through these lines if:

(a) a red filter is placed in front of the tungsten lamp,
(b) a blue filter is placed in front of the tungsten lamp,
(c) no filter is used.
(d) What happens to the observed fringes if the lamp with the blue filter in front of it is moved further away?

123 A student wishes to observe an interference pattern by looking through a double slit at a source of red light ($\lambda = 600$ nm). She stands 2 m away from the source and the slits are narrow and 1 cm apart. Describe, with numerical details, what she sees and comment on an improvement which could be made.

124 A diffraction grating is required which will produce a high degree of dispersion.

(a) What property of the grating is important to meet this requirement?
(b) When a white light source is viewed through this grating a central white band is obtained with a spectrum on either side. Discuss in terms of the wave nature of light whether the blue or the red end of the spectrum is closest to the white band. Explain the presence of the white band.

Heat and the kinetic theory of gases

Relevant values of specific heat capacities, specific latent heats, and densities may be found from the science data booklet.

125 A waterfall 300 m high has water flowing over the top at a rate of 100 kg per second. Calculate the rise in temperature of the water at the bottom. State any assumptions that have been made. How much power is available?

126 A lead bullet of mass 20 g travelling at 200 m s^{-1} comes to rest in a block of naphthalene of mass 50 g. If the initial temperatures of the bullet and naphthalene are both 20°C, find:

(a) the kinetic energy of the bullet just before entering the naphthalene,
(b) the final temperature of the naphthalene and bullet. State any assumptions made.

127 A 30 g block of ice at 0°C was placed in a polystyrene cup containing water at 25°C. The final temperature was found to be 5°C. How much water was in the cup? State any assumptions made.

128 Water flows at a steady rate of 0.04 kg s^{-1} through a glass tube containing a 650 W heating element. If the initial temperature of the water entering the tube is 18.2°C and the temperature of the water leaving the tube is 21.8°C:

(a) calculate the heat loss per second from the sides of the tube,
(b) How could the heat loss be reduced?

129 The graph below shows the variation of temperature with time when a 0.45 kg block of solid metal at 80° is heated at a constant rate of 75 W until it has all melted.

(a) Calculate the specific heat capacity of the metal.
(b) Calculate the specific latent heat of fusion.
(c) What metal do you think has been used?

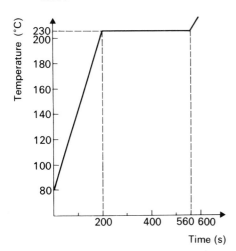

130 In an experiment to determine the specific heat capacity of a solid, a block of the solid is suspended in a beaker of boiling water for a few minutes, then transferred into another beaker containing a measured amount of water (*m*) at 20°C. The final temperature of the water (*T*) is recorded.

(a) What other measurements are required?
ʲ(b) Explain in general terms how the specific heat capacity is calculated from these results.
(c) Comment on the main sources of error in this experiment and describe what you would do to reduce them to a minimum.

131 The 10 kg mass hangs at a constant height above the floor. The initial reading of the thermometer inside the cylinder is 17°C. After the handle is turned 120 times, the temperature is observed to rise to 25°C. If the circumference of the cylinder = 0.1 m and the mass of the cylinder = 0.3 kg,

(a) calculate the specific heat capacity of the metal, stating any assumptions made.

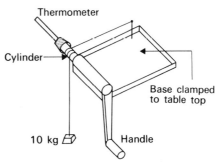

(b) Another pupil repeats the experiment with a 5 kg mass but produces the same temperature rise. How many times did he turn the handle? Do you think he took longer to complete the experiment?
(c) Discuss the main sources of error and how you would minimize them.

132 A small electrical heater, rated at 10 W, was placed in a block of ice for 7 minutes, and it was found that 11 g of ice melted.

(a) Determine the apparent efficiency of the heater.
(b) Comment on any assumptions that you have made.

133 Radium-224 decays into the gas radon. If the half life of radium is 3.5 days how many atoms of radon are produced from a 1 mg sample of radium in 10.5 days?

134 A 3-litre vessel of oxygen at 10^5 Pa is joined by a connecting tap to a 4 litre vessel of nitrogen at 2×10^5 Pa.

(a) Determine the pressure of the mixture.
(b) If the initial temperature of both gases was 17°C, what would be the pressure of the mixture if heated to 91°C?

135   In the experiment shown, the atmospheric pressure is 75.6 cm of mercury.

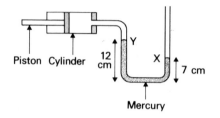

Piston   Cylinder   12 cm   X   Y   7 cm

Mercury

(a) Find the pressure of the gas inside the cylinder.
(b) If the piston is pushed inwards such that the volume decreases by 10%, estimate values for the heights X and Y to the nearest cm.
(c) What would you expect to happen to the mercury levels X and Y if the cylinder was gently heated?

136   A tray of water is lightly dusted with lycopodium powder. A small drop of oil of diameter 1.2 mm is placed gently in the centre of the water surface where it spreads out to cover a circular area of diameter 60 cm.

(a) Estimate the length of the oil molecule, stating any assumptions made.
(b) That precautions should be taken with the water surface?
(c) Why is lycopodium powder sprinkled on the water surface?

137   A rectangular block of mass 3 kg has a base 10 cm by 10 cm and height 40 cm, and is stationary on a table.

(a) State the pressure on the table under the block.
(b) What is the pressure if the block is placed on its side?

138   A gas cylinder has a volume of 6 litres and contains oxygen at a pressure of 10 atmospheres. How much oxygen is available at standard pressure? State any assumptions you have made.
(1 atmosphere = 10^5 Pa)

139   According to the kinetic theory of gases, pressure and volume are related by the expression:

$$pV = \tfrac{1}{3}Nm\overline{c^2}.$$

(a) Explain the significance of the fraction $\tfrac{1}{3}$ and the term $\overline{c^2}$.
(b) Use the above expression to derive a formula for the root mean square velocity of a gas in terms of the pressure and density.
(c) From this formula determine the root mean square velocity of argon molecules at s.t.p.
(d) How would this r.m.s. velocity be affected if:

  (i) the temperature was increased without a change in pressure?
  (ii) the pressure was increased without a change in temperature?

140   A relationship between the pressure (p) and volume (V) of an ideal gas is given by the kinetic theory:

$$pV = \tfrac{1}{3}N\,m\overline{c^2}.$$

(a) Show that this question is consistent with the gas law, $\dfrac{pV}{T}$ is a constant for a fixed mass of gas; (or $pV \propto T$ if the mass is constant).
(b) A certain mass of gas is contained in a cylinder and gently heated for a few minutes; some gas is allowed then to escape until the pressure returns to its initial value. Consider each quantity p, N, m, $\overline{c^2}$ separately and state if its value has changed: (i) during, (ii) at the end, of this experiment.

141
(a) In the kinetic theory of gases, use is made of the relationship

$$T \propto \tfrac{1}{2}m\overline{c^2}.$$

What physical quantity is associated with the term $\tfrac{1}{2}m\overline{c^2}$?
(b) Show that the ratio of the root mean square velocities of carbon dioxide (CO_2) to hydrogen (H_2) molecules is $1:\sqrt{22}$, if both gases are at 20°C.

(c) Why does this help to explain the lack of hydrogen in the Earth's atmosphere?

(d) If the root mean square velocity of hydrogen (H_2) is $1900\,m\,s^{-1}$ at 20°C, at what temperature does helium (He) have the same root mean square velocity?

Relative atomic masses: $H = 1$, $He = 4$, $C = 12$, $O = 16$.

142 A certain volume of helium at s.t.p. contains 10^{24} molecules. How many oxygen molecules would this same volume contain at s.t.p.? Which law does your answer illustrate?

143 A statement in a book reads, 'The molar volume of a gas at s.t.p. is $22.4 \times 10^{-3}\,m^3$.' Explain clearly what is meant by this statement.

Another statement reads, 'Avogadro's constant $= 6 \times 10^{23}\,mol^{-1}$.' Explain in detail the physical significance of the number 6×10^{23} and the unit—the mole.

144 From the kinetic theory of gases, the relationship $pV = \frac{2}{3}N(\frac{1}{2}m\overline{c^2})$ may be derived.

(a) Show how this equation predicts Charles' law relating the volume and temperature of a gas.

(b) Describe an experiment which might verify this law experimentally.

(c) What is meant by an ideal gas?

(d) Under what conditions would you expect a gas *not* to obey the ideal gas law, namely $\dfrac{pV}{T}$ is a constant for a fixed mass of gas?

145 Zinc has a density of $7.13 \times 10^3\,kg$ m^{-3} and a relative atomic mass of 65. Find:

(a) the mass of one atom of zinc in kg,

(b) the volume of one atom and hence estimate the distance between the centres of the zinc atoms in the solid state,

(c) the approximate separation of the atoms in the gaseous state.

146

(a) State the difference between an ideal gas and a real gas.

(b) Describe the difference in structure between liquids and gases.

(c) State why the kinetic theory of gases cannot be applied to liquids or solids.

The atom and the nucleus

147 A prism spectrometer is set up to view the emission spectrum from a helium vapour lamp.

(a) Explain *why* a series of coloured lines are observed.

(b) Is there any change when a red filter is placed between the collimator and the prism?

(c) The yellow line in this helium spectrum has a wavelength of 588 nm. Calculate the frequency of this radiation and the energy of the photons.

(d) Explain carefully the difference between the emission spectrum and the absorption spectrum of sodium vapour.

148 A clean zinc plate is fitted to the top of a negatively charged gold leaf electroscope and illuminated by ultra-violet radiation for 5 seconds.

(a) Explain carefully why the leaf is observed to fall.

(b) What difference would you expect if:

(i) the intensity of the ultra-violet is increased?

(ii) the ultra-violet source is replaced by a green/blue light?

(iii) the electroscope had been given a positive charge?

149 Two coherent beams of light are required to demonstrate interference. With reference to atomic energy levels, explain what is meant by coherence.

150 Explain why the photoelectric effect and the emission spectrum of an element illustrate the particulate nature of light. What is the name given to a particle of light? Explain the term quantum.

151 With reference to Figure 7.1, page 205, state the ionization energy of hydrogen. Briefly explain what is meant by the term ionization.

152 Briefly comment on the experimental evidence for the wave–particle duality of electrons and electromagnetic radiation.

153
(a) Explain what is meant by the half-life of a radioactive isotope.
(b) An impure radioactive sample is found to emit α and β particles. Describe in detail how you would measure the half-life of the β emission. Draw a clear diagram showing the experimental arrangement.
(c) In an experiment to determine the half-life of the isotope iodine-131 the following results were recorded:

background count = 21 counts/minute

time (days)	0	3	6	9	12	15
counts/minute	107	88	73	61	53	46

Determine by graphical means the half-life of this isotope.

(d) The radioactive decay of iodine-131 may be written;

$$^{131}_{53}I \xrightarrow{\ \beta\ } {}^{A}_{Z}X$$

What are X, A, and Z?

154 In β decay an electron is emitted from the nucleus. Explain how an electron can leave a nucleus which only contains protons and neutrons.

155 Certain physical quantities are said to be conserved.

(a) State the conservation laws associated with the following quantities: (i) energy, (ii) momentum, (iii) charge.
(b) Are these quantities vectors or scalars?
(c) Give examples to illustrate each of the conservation laws.

156 In the Rutherford scattering experiment performed by Geiger and Marsden, about 1 in 8000 particles were deflected backwards. Explain why these events led to a new model of the atom.

157 With reference to beryllium $^{9}_{4}Be$, explain what is meant by the mass defect and binding energy of a nuclide.

158 With reference to the reaction

$$^{1}_{0}n + {}^{235}_{92}U \longrightarrow {}^{92}_{36}Kr + {}^{141}_{56}Ba + 3{}^{1}_{0}n$$

(a) Explain what is meant by a chain reaction.
(b) Explain why this reaction is termed fission.
(c) Is the mass of the neutron plus uranium equal to the mass of the products? Explain what happens to any 'lost' mass.

159 Distinguish between *fusion* and *fission*, and give an example of each.

160 Rewrite the following reactions using chemical symbols for the nuclides and give atomic numbers and mass numbers for *all* the constituent nuclides, e.g. $^{1}_{1}H$ for the proton.

(a) $^{214}Po \longrightarrow X + \alpha$.
(b) $^{23}_{11}Na + n \longrightarrow {}^{24}Na + X$.
(c) $4p \longrightarrow \alpha + 2X$. (This is an overall result of a sequence of reactions involving deuterium and tritium.)

Answers to the Exercise Section

1 Distance $= 16$ km; displacement $= 7.9$ km due NW.

2 (a) No (b) 30 N, bisecting 120° angle.

3 (a) $1300 \, \mathrm{m\,s^{-1}}$ at 22.6° N of W, bearing (293°) (b) At $1200 \, \mathrm{m\,s^{-1}}$ 24.6° S of W, or bearing (245°).

4 (a) $25 \, \mathrm{km\,h^{-1}}$, 16.3° W of N (b) $25 \, \mathrm{km\,h^{-1}}$, 16.3° E of N.

5 $3.25 \, \mathrm{m\,s^{-1}}$, 22.6° W of N; time 4 s.

6 (a) $3 \, \mathrm{m\,s^{-2}}$ (b) 33 m (c) $4\frac{1}{8} \, \mathrm{m\,s^{-1}}$

(d)

t	1	2	3	4	5	6	7	8
s	1.5	6	12	18	24	29	32	33

7 30 m, falling downwards.

9 (b) 0–75 s, $v = 4 \, \mathrm{m\,s^{-1}}$; 75–78 s, $v = 0$; 78–168 s, $v = -3.3 \, \mathrm{m\,s^{-1}}$. Assumptions: velocity constant in each range, accelerations neglected.

10 (a) No (b) $+2 \, \mathrm{m\,s^{-2}}$, $-3 \, \mathrm{m\,s^{-2}}$, $+1.5 \, \mathrm{m\,s^{-2}}$.
(d) 0–10 s, $a = 2 \, \mathrm{m\,s^{-2}}$; 10–30 s, $a = 0$; 30–45 s, $a = -3 \, \mathrm{m\,s^{-2}}$; 45–55 s, $a = 1.5 \, \mathrm{m\,s^{-2}}$.

11 (a) $7 \, \mathrm{m\,s^{-1}}$ (b) $9 \, \mathrm{m\,s^{-1}}$ (c) 25 m.

12 Distance $= 162.5$ m, displacement $= 140$ m.

13 (a) $17.5 \, \mathrm{m\,s^{-2}}$, $25 \, \mathrm{m\,s^{-2}}$, F/t graph—line parallel to x-axis for 60 s then zero, a/t graph—line inclined to x-axis from $17.5 \, \mathrm{m\,s^{-2}}$ up to $25 \, \mathrm{m\,s^{-2}}$ at 60 s then zero (b) a curve (c) area under a/t graph $= 1275 \, \mathrm{m\,s^{-1}}$.

14 (a) 0.5 s (b) $20.6 \, \mathrm{m\,s^{-1}}$, 14° to horizontal.

15 (a) Position: 20.8 m horizontal, 7 m vertical. Velocities: $20.8 \, \mathrm{m\,s^{-1}}$ horizontal, $2 \, \mathrm{m\,s^{-1}}$ vertical. (b) 7.2 m (c) 49.9 m, 2.4 s.

16 (a) 125 m (b) 10 s (c) 866 m.

17 10 m.

18 (a) $9.9 \, \mathrm{m\,s^{-2}}$ (d) less than $9.9 \, \mathrm{m\,s^{-2}}$.

19 (a) 25 flashes per second (b) $3.25 \, \mathrm{m\,s^{-1}}$.

21 (a) $26 \, \mathrm{m\,s^{-1}}$ (b) $39.7 \, \mathrm{m\,s^{-1}}$ at 49° to horizontal.

22 Initial acceleration is $1 \, \mathrm{m\,s^{-2}}$ but acceleration will increase as the chain slides off the bench.

23 (a) 4 s (b) $0.707 \, \mathrm{m\,s^{-1}}$ towards the centre.

24 (a) 49 N (b) 8 N (c) 18.5 N (d) 44.5 N.

25 (a) $2 \, \mathrm{N\,kg^{-1}}$ (b) 16 m, $11.3 \, \mathrm{m\,s^{-1}}$.

26 (a) $2.45 \, \mathrm{m\,s^{-1}}$.

27 18 m from the start travelling at $3 \, \mathrm{m\,s^{-1}}$.

28 (a) 8.48 N, $7.07 \, \mathrm{m\,s^{-2}}$ (b) 6 N, $5 \, \mathrm{m\,s^{-2}}$.
(c) component of weight unchanged, accelerations reduced to (a) $6.2 \, \mathrm{m\,s^{-2}}$, (b) $4.2 \, \mathrm{m\,s^{-2}}$.

29 (a) $0.4 \, \mathrm{m\,s^{-2}}$ (b) $1.0 \, \mathrm{m\,s^{-1}}$ (c) $2.0 \, \mathrm{m\,s^{-1}}$ (d) $2 \, \mathrm{m\,s^{-2}}$ (e) 0.8 N.

30 (a) 1.6 m (b) 1.3 m.

31 34×10^3 N.

32 (a) $1 \, \mathrm{m\,s^{-2}}$ (b) $0.8 \, \mathrm{m\,s^{-2}}$.

33 (a) 20 N (b) 120 W (c) $1.5 \, \mathrm{m\,s^{-1}}$.

34 (a) 200 N (b) 833 N (c) 133 N.

35 (a) 15 J (b) 3 J (c) 12 J (d) 12 J (e) $4 \, \mathrm{m\,s^{-1}}$.

37 (a) 0.06 J (b) $2 \, \mathrm{m\,s^{-1}}$ (d) 0 (e) 0.4 s, 2.5 Hz.

38 (a) Steady speed up *or* down (b) Acceleration of $1\,\mathrm{m\,s^{-2}}$ downwards *or* deceleration of $1\,\mathrm{m\,s^{-2}}$ upwards (c) acceleration of $2\,\mathrm{m\,s^{-2}}$ upwards *or* deceleration of $2\,\mathrm{m\,s^{-2}}$ downwards.

39 (a) $0.06\,\mathrm{kg\,m\,s^{-1}}$ (b) $0.4\,\mathrm{s}$ (c) $0.8\,\mathrm{m}$ out from base of table (d) horizontally $2\,\mathrm{m\,s^{-1}}$, vertically $4\,\mathrm{m\,s^{-1}}$ (e) $0.3\,\mathrm{J}$.

41 (a) $4\,\mathrm{m\,s^{-1}}$ (b) $0.16\,\mathrm{kg\,m\,s^{-1}}$ (c) $0.32\,\mathrm{J}$ (d) $0.64\,\mathrm{J}$ (e) $0.23\,\mathrm{kg\,m\,s^{-1}}$.

42 (a) $0.21\,\mathrm{kg\,m\,s^{-1}}$ (b) $0.21\,\mathrm{N\,s}$ →∫

(c) $0.21\,\mathrm{N\,s}$ ←∫ (d) $7 \times 10^2\,\mathrm{N}$.

43 75%.

44 F increases to a maximum of $500\,\mathrm{N}$ at $1.2\,\mathrm{ms}$ then decreases to zero at $2.4\,\mathrm{ms}$, area under graph $= 0.6\,\mathrm{N\,s}$.

45 $2.7\,\mathrm{m\,s^{-1}}$.

46 (a) $2\,\mathrm{J}$, $1.8\,\mathrm{J}$ (b) $1.23\,\mathrm{kg\,m\,s^{-1}}$ (c) $1.23\,\mathrm{N\,s}$ (d) $411\,\mathrm{N}$.

47 The time for the ball to interrupt the light beam $= 0.02\,\mathrm{s}$.

48 (a) $5\,\mathrm{m\,s^{-1}}$ (b) $750\,\mathrm{N}$.

49 (b) $20\,\mathrm{V}$, $0.06\,\mathrm{N}$.

50 (b) $2.05 \times 10^7\,\mathrm{m\,s^{-1}}$ (c) $7.0 \times 10^{14}\,\mathrm{m\,s^{-2}}$ perpendicular to Y plates. (d) $6.8 \times 10^5\,\mathrm{m\,s^{-1}}$ (e) $2°$ (f) $10\,\mathrm{mm}$, which is small.

52 (a) $2\,\mathrm{kV}$ (b) $5 \times 10^{-9}\,\mathrm{C}$, upthrust of air neglected (c) 3.1×10^{10}.

53 6.

54 (a) $6.4 \times 10^{-15}\,\mathrm{N}$ for both (i) and (ii) (b) $6.4 \times 10^{-16}\,\mathrm{J}$ (c) $3.75 \times 10^7\,\mathrm{m\,s^{-1}}$ (d) $3.75 \times 10^7\,\mathrm{m\,s^{-1}}$ (e) $6.4\,\mathrm{W}$.

55 1.25×10^{19} electrons.

56 $1:0.5:2$.

57 (a) $8.5\,\Omega$ (b) $4\,\mathrm{A}$, $2\,\mathrm{A}$ (c) $34\,\mathrm{V}$.

58 $60\,\Omega$, $30\,\Omega$, $13.3\,\Omega$, $6.7\,\Omega$.

59 (a) $5.6\,\Omega$, $6\,\Omega$, $6.9\,\Omega$, $8.1\,\Omega$, $8.3\,\Omega$.

60 (a) $0.5\,\mathrm{A}$ (b) $1\,\mathrm{V}$ (c) $+1\,\mathrm{V}$ (d) $3\,\mathrm{W}$.

61 (a) $0.6\,\mathrm{A}$ (b) $2.4\,\mathrm{V}$, $1.2\,\mathrm{V}$ (c) $0.8\,\mathrm{V}$.

62 (a) (i) $20\,\mathrm{V}$, (ii) $20\,\mathrm{V}$ (b) (i) $16\,\mathrm{V}$, (ii) $20\,\mathrm{V}$ (c) currents through $1\,\Omega$, $4\,\Omega$ and $10\,\Omega$ are (i) $4\,\mathrm{A}$, $4\,\mathrm{A}$, $2\,\mathrm{A}$, (ii) $0.5\,\mathrm{A}$, $2\,\mathrm{A}$, respectively.

63 $120\,\mathrm{W}$.

64 (a) $8\,\mathrm{V}$, $3\,\Omega$ (b) $17\,\Omega$ (c) $3.2\,\mathrm{W}$, $2.72\,\mathrm{W}$ (d) when $R = 3\,\Omega$ power $= 5.3\,\mathrm{W}$ (e) when $R = 99\,\Omega$, p.d. across $R = 7.76\,\mathrm{V}$.

65 (a) $0.8\,\mathrm{V\,\Omega^{-1}}$ (b) $0.4\,\mathrm{V\,cm^{-1}}$ (c) $35\,\mathrm{cm}$.

66 (a) Shunt $0.1\,\Omega$ (b) multiplier $4950\,\Omega$ (c) sensitivity per division of (a) $0.01\,\mathrm{A}$, of (b) $0.1\,\mathrm{V}$.

67 (a) $960\,\Omega$ (b) $4\,\mathrm{V}$ (c) $3\,\mathrm{V}$ (g) decreases.

68 (a) S_1, S_2 both closed, $R_1 = 0.05\,\Omega$ (b) S_1, S_2 both open, $R_2 = 49\,500\,\Omega$.

70 (c) $5\,\Omega$.

71 $1200\,\Omega$, $400\,\Omega$, $300\,\Omega$, $100\,\Omega$ resistors used in the bridge, $600\,\Omega$ resistor and switch used to protect the galvanometer.

72 (a) $37.5\,\mathrm{cm}$ (b) (i) $3\,\mathrm{mA}$ (ii) $1.5\,\mathrm{mA}$ in reverse direction.

73 No current through R_3; if diodes are identical, currents through R_1 and R_2 are the same.

74 (a) $A_1 > A_2$ (b) reading same as V_s, transistor is off (c) both zero (d) n-p-n.

75 $V_{\mathrm{r.m.s.}}$ across $20\,\Omega$ is $40\,\mathrm{V}$, peak p.d. is $56.6\,\mathrm{V}$.

77 To check that the power is the same.

78 Peak voltage $= 42.4\,\mathrm{V}$, peak-to-peak voltage $= 84.8\,\mathrm{V}$.

79 $4\,\mathrm{cm}$.

82 $1.5\,\mathrm{mC}$, $4.5\,\mathrm{mJ}$.

83 (a) leaf rises (b) leaf falls.

85 (a) $1.5\,\mathrm{mC}$ (b) $15\,\mu\mathrm{F}$ (c) both double.

87 (b) $0.2\,\mathrm{mA}$, $20\,\mathrm{V}$ (c) (i) take longer to charge C (ii) C charged in shorter time.

88 (a) $800\,\mathrm{kW}$ (b) $320\,\mathrm{W}$, transformer 100% efficient.

89 (a) $2.5\,\mathrm{V}$ (b) $0.9\,\mathrm{A}$ (c) 90% (d) primary current $= 0.0167\,\mathrm{A}$.

90 (a) full wave rectified trace, peak value $21.2\,\mathrm{V}$ (b) zero p.d.

91 (a) Low, prevent heat loss (b) $0.14\,\Omega\,\mathrm{km^{-1}}$ (c) steel for support (d) $550:1$ step down (e) transformer does not function with d.c.

95 (a) An inductor, back e.m.f. at break (b) a capacitor (c) for Z increase the rate of vibration of S, for Y reduce value of C.

96 (c) only reactance varies with frequency.

97 (a) peak value $2.12\,\mathrm{V}$ (b) same (c) (i) Y_1 trace has the smaller amplitude, (ii) Y_1 trace has the larger amplitude (d) p.d. (e) high impedance.

98 Both graphs are straight lines, (*b*) has the larger gradient.

99 The graphs are curves, *I* decreases as *f* increases, (*a*) lies below (*b*).

100 (*a*) R_2 (*b*) R_1.

101 An inductor chokes a.c. variations, a capacitor blocks direct current.

103 26.3°.

104 (*a*) 49.7° (*b*) 70.1° (*c*) total internal reflection.

105 1.41.

106 36°.

107 Total internal reflection at the sides.

108 (*a*) Real inverted diminished, $m = 0.67$ (*b*) image size increases (*d*) magnification increases $m = 2$ at $u = 12\,cm$.

109 (*a*) image is real inverted, 37.5 cm from lens, magnification = 1.5 (*b*) image is real inverted, 183 cm from the lens, magnification = 7.3.

110 (*a*) 0.7 m.

113 $\lambda = 398\,nm$, $f = 4.74 \times 10^{14}\,Hz$, $v = 1.89 \times 10^8\,m\,s^{-1}$.

114 0.125 m.

115 (*b*) No, E varies as $1/d^2$ (*c*) 88.2 units.

117 (*a*) 8 s (*b*) 1 cm.

119 (*a*) λ (red) $> \lambda$ (blue) (*b*) No λ (blue) is reduced more.

121 (*a*) 10 cm (*b*) 3300 Hz (*c*) 40 cm.

123 Fringe separation = 0.12 mm. This is very small. To improve the fringe separation, reduce the slit separation.

125 0.7°C, 300 kW.

126 25.7°C.

127 127 g.

128 48 J.

129 (*a*) $222\,J\,kg^{-1}\,K^{-1}$ (*b*) $6 \times 10^4\,J\,kg^{-1}$.

131 $500\,J\,kg^{-1}\,K^{-1}$.

132 87.5%.

133 2.3×10^{18} atoms.

134 (*a*) $1.57 \times 10^5\,Pa$ (*b*) $1.97 \times 10^5\,Pa$.

135 (*a*) 70.6 cm Hg (*b*) Y = 8 cm, X = 11 cm (*c*) Y decreases.

136 3.2 nm, monomolecular layer.

137 (*a*) 3 kPa (*b*) 750 Pa.

138 54 litres, constant temperature.

139 (*c*) $410\,m\,s^{-1}$.

141 586 K (313°C).

142 10^{24}, Avogadro's law.

145 (*a*) $1.08 \times 10^{-25}\,kg$ (*b*) $1.51 \times 10^{-29}\,m^3$, separation 0.25 nm (*c*) approx. 2.5 nm.

147 (*c*) $5.1 \times 10^{14}\,Hz$, $3.4 \times 10^{-19}\,J$.

151 $21.8 \times 10^{-19}\,J$.

153 (*c*) 8.1 days (*d*) $^{131}_{54}Xe$.

154 $n \longrightarrow p^+ + e^-$.

157 mass defect = mass of $(4p + 5n)$ − mass of ^9_4Be.

160 (*a*) $^{214}_{84}Po \longrightarrow ^{210}_{82}Pb + ^4_2He$
(*b*) $^{23}_{11}Na + ^1_0n \longrightarrow ^{24}_{11}Na + ^0_0\gamma$
(*c*) $4^1_1H \longrightarrow ^4_2He + 2^0_1e$ (^0_1e is the positron).

Bibliography

Books designed specifically for S.C.E. Higher Grade Physics include:

Scottish Certificate of Education, Past Papers in Physics (*Higher Grade*) (Glasgow: Gibson, published annually).

Upgrade your Physics, R. Neill and G. Sydserff (Edward Arnold, 1980).

H Grade Questions in Physics, Lanarkshire Physics Group (London: Heinemann Educational, 1972).

H Grade Physics, S. G. Burns (London: English Universities Press, 1974).

Higher Physics, J. Jardine (London: Heinemann Educational, 1983).

The following text is most useful for the linking of computers to experiments:

Microcomputers and Science Teaching, R. A. Sparkes (London: Hutchinson, 1982).

Texts which will be useful for providing descriptions of experiments and further background material include:

Nuffield Guide to Experiments, Nuffield Foundation (Harlow: Longman/ Penguin, 1966).

PSSC Physics (Farnborough: D.C. Heath and Co., 1965).

Physics for the Enquiring Mind, E. M. Rogers (Princeton, New Jersey: Princeton University Press, 1960).

Advanced Level Physics, 5th Edition, M. Nelkon and P. Parker (London: Heinemann Educational, 1982).

Patterns in Physics, W. Bolton (Maidenhead: McGraw-Hill, 1974).

Electricity and Atomic Physics, R. Brown (London: Macmillan, 1973).

The Scottish Curriculum Development Service, Dundee Centre, issue useful memoranda describing the optional Special Topics. These include:

Memorandum No. 54 *Nuclear Reactors* (This booklet contains excellent pupil material).

Memorandum No. 52 *Optical Instruments*.

Index